Undaunted

The Reunion of the Fighting Men

A Narrative from the First Battalion Fifth Marines

Undaunted is a narrative from the men of the First Battalion Fifth Marines of the United States Marine Corps in the years of 2003-2005, followed by an an account of what many of these men have experienced after returning to civilian life. The author has done her best to ensure that the narrative is accurate, spending many hours listening to the Marines and their families, then relistening to the recorded interviews with final review from each interviewee. Any differences in the author's narrative and what others may remember is not intentional or intended to harm others. The author and publisher and all associated with this book, directly or indirectly, disclaim any liability, damage, loss or injury resulting from this account.

Undaunted: The Reunion of the Fighting Men by Ruth Mortenson
First Edition 2025
© Ruth Mortenson
Text All Rights reserved to Ruth and Ken Mortenson
Cover and Interior Photos credit: Al Alcala-Esperanza, © 2025
Back Cover Art: Tony Reed, © 2025

Publishing by Morten Moore Publishers
415 E. Mohawk
Flagstaff, AZ 86005

ISBN 979-8-9995506-0-6

Undaunted

Contents

Preface

Chapter 1	The Invitation	1
Chapter 2	The Call to Serve	5
Chapter 3	The Fighting Fifth	11
Chapter 4	Muster	31
Chapter 5	Undaunted	43
Chapter 6	1st Sgt's Hill	53
Chapter 7	A Measure of Respect	61
Chapter 8	To Command	75
Chapter 9	Corpsman Up!	87
Chapter 10	Families of the Fallen	99
Chapter 11	Dining with Robert Irvine	119
Chapter 12	Gold Star Families	125
Chapter 13	Sunday Morning	137
Chapter 14	Homeward Bound	145
Chapter 15	Reflections	155

Dedicated to
The Marines of the First Battalion Fifth Marines

I want to thank each of the Marines and family members that spoke with me regarding 1/5's tours in Iraq and the events surrounding this conflict. I could not have written this book without their openness. Thank you for talking with me so freely and for reviewing what was written even though the re-telling brought difficult memories to the surface.

Preface

A Marine Battalion contains three to five companies of men, totaling around 800 to 1200 individuals. Each has his own story. The men of the First Battalion of the Fifth Marines say that their stories will never go untold.

As I began to interview, research and write *Undaunted*, I humbly admitted that I could not possibly understand what the Marines of the 1/5 endured or the struggles some have faced. I believe their stories should be told with the hope that their underlying resilience will shine through and encourage others.

One book cannot contain all their stories. How could I select men that would best represent their brethren without slighting others? I relied on Major General Fred Padilla, Sergeant Major Luke Converse and the men I met at the reunion to bring individuals to my attention. Each of these men share common elements within their stories that represent the men of the 1/5.

The manuscript is not intended to recount their deeds in the battles that took place in Iraq from 2003 through 2005. Rather, I wish to look at how these men are faring 20 years after combat. The word trauma has been overused in recent years. These men have seen their friends blown up or been caught in explosions that have damaged their brains and bodies. Their struggle was born in the act of killing before they, in turn, were killed. I think back to the men of WWII and the Korean Conflict. Few of those who had seen combat readily talked about their experiences. The same is true of many of the men who fought in Vietnam. They remained silent, knowing the listener might not identify with the violence they experienced. Today, the Marines of the 1/5 follow the same pattern, rarely speaking to those who have not served of what they lived in Iraq and Afghanistan, in Africa and eastern Europe.

Their nightmares come, disturbing others. These men react to

First Battalion fifth Marines in front of Saddam Hussein's palace.

sudden, loud noises. Depression may seep into the cells of their brains. These are lonely moments for the survivors of combat.

Come, sit awhile and consider their stories. They want you to know what they accomplished, what they suffered on behalf of our country and the countries held captive by despots. Some hesitate to say they accomplished anything of worth but I can assure you that in the moment, they were valiant, fighting on behalf of those held captive by greed and violence. Let them tell their stories. And listen without judgement, without pity, accepting their sacrifice.

I have asked a number of readers from the 1/5 as well as military service to read this manuscript and make corrections. We have made every attempt to correct any errors and typos. Any errors are mine, not out of avarice but simply misunderstanding what has been said and done. Mea culpa!

To the men of the First Battalion of the Fifth Marine Regiment , thank you for speaking with me and for revealing your quiet struggles. Most of all, thank you for your willingness to stand watch over our country and take up the battle for freedom on the behalf of others.

Chapter 1

The Invitation

Reunions come in all sizes. Our invitation arrived in our mail box. The invitation was unusual. Rarely are the families of those killed in combat united with their son's former comrades-in-arms.

"January of 2023 marks the twentieth anniversary of 1st Battalion, 5th Marines', deployment to Kuwait and subsequent invasion into Iraq for Operation Iraqi Freedom (OIF). It is my honor to have served as battalion commander during this first of three rotations of the battalion for combat operations in Iraq.

"The Marines of the battalion are planning a reunion to commemorate the anniversary of the first deployment of the battalion to Iraq. With unanimous consent, the committee has decided to include all three rotations (2003, 2004 & 2005) of OIF in this reunion. As you know, the Marines and Sailors of the battalion served with bravery and distinction. Their service and sacrifice are part of a legacy that should be honored and celebrated. This reunion will serve to remember your Marty's sacrifice and service as well keep the flame of camaraderie burning bright.

"It is my privilege to extend to you a personal invitation on behalf of the 1st Battalion, 5th Marines, Reunion Committee. As a Gold star Family, you are an honored part of our legacy that holds a special place in the hearts of all who have served with this battalion. I hope you will be able to attend and share in this experience as we go back to Camp Pendleton for this reunion during 7-9 April 2023. As honored guests expenses associated with your travel, lodging and attendance at all reunion events will be covered."   ~ Major General Fred Padilla, Ret.

Stepping into the elevator of the Marriot in Oceanside, I pressed the button for the ground floor. Descending to the opening reception, my emotions were mixed between apprehension and anticipation in meeting the men who had served with my son. Years earlier, upon their return from Iraq after their third deployment, we had been present when the battalion strode onto the Parade Deck in formation, their steps in unison against the pavement.

As the ceremony ended, the men were released from formation. Along with the families, I had surged into the crowd of uniforms seeking the corpsman that had treated my son when he was fatally wounded. He glared at me, a Marine to either side, both scowling.

"What do you want?" he demanded.

I wasn't the enemy. Why was he glaring at me?

Later, over a company-wide dinner we had been mostly ignored until the final moments when their officers ordered the men to file past, shaking our hands. The experience was uncomfortable. Many of the men were in feeling survivor's guilt, grieving over those lost. They covered their grief with loud voices, bluster and bravado. Each man knew he had won the lottery, returning home in one piece. Extending their hands, meeting our eyes, they muttered, "Sorry for your loss." Out of the line-up, four quietly told us, "Your son saved my life."

And then, they moved forward, anticipating their return to civilian life.

Twenty years later, I moved through the crowd, searching unsuccessfully for two, maybe three faces we knew from previous contact. The reunion flowed from the conference room out into a patio, the atmosphere relaxed as families and single men caught up on the previous 20 years. The battalion consisted of four companies: Alpha, Bravo, Charlie and Weapons. * Some men successfully transitioned into society while others struggled with their memories, choosing to muffle their pain with alcohol. Some, unable to silence the of violence in their brains, could not attend as they lay in quiet graves in their home towns.

Each of the veterans who attended wore a name tag with their name and company. As I moved through the crowd, I focused on

* The names or call signs for these companies change with each deployment.

name tags, looking for members of Alpha as those men were most likely to have served with my son. Some of the veterans were accompanied by their families, each man wore his experience as a testament to his service on behalf of our country. Intermingled with the Marines were families missing their sons and husbands. Name tags dangling from Gold ribbons signified the Gold Star families who had lost their men in combat. Others wore a tag with a black ribbon signifying a man lost after he had re-entered civilian life.

Therein lies the story. Walk with me as I introduce you to some of the men who once faced the violence and camaraderie of war and consider how each man moved through the challenge of re-entering society. The discussion that follows is not meant to give another recital of the conflict in Iraq but instead relates to where the men of the 1/5 are twenty years after returning home. We might wish that each has gone on to lead a successful life but that has not happened for some of the veterans. Their stories are painful but they are honest. Our country owes it to these men to listen to their transition in becoming the men they are today.

Chapter 2

The Call to Serve

When a man enlists in the United States Marine Corps, he is sent
to an intake center. The Marine Corps begins by removing what
is familiar to the recruit. Heads are shaved, personal possessions
sent back home, civilian clothes are locked away and uniforms
dispensed. Marines are not peacekeepers, given to keeping law
and order. They are sent into some of the toughest conflicts in the
world, to defeat those opposed to the rule of civil law. They describe
their role as 'breaking things' before handing the responsibility
of establishing a new order over to the Army. To complete their
missions, these men must be transformed from individuals with
independent goals to a unified fighting force. The weeks spent
in boot camp are intended to detach the man from what he has
known, bring him into top physical shape and rebuild the recruit to
the standards of the Corps first established in July 1798. Once boot
camp is completed the Marine will either be assigned to infantry or
he will enter training for a specific job within the Corps.

Eric Young was one such young recruit. He arrived at the
Marine Corps Recruitment Depot fresh out of high school. He
describes his upbringing as rocky. His mom left his dad, taking
their two sons when Eric was four years of age. He loved spending time with both sets of grandparents even as he missed having
his dad around. He wasn't all that interested in school, a common
refrain from many of the Marines. At age eight, he met his stepfather. In the years that followed, they had a contentious relationship
and Eric was eager to leave home, getting away from his stepdad's
restrictions.

"My step-dad didn't know how to be a father," he recalls.

"I think he did the best that he knew but he used restriction to discipline me. He would take my stuff when I would step out of line. Years later, I found a whole stash that he never got around to returning."

He intended to join the Air Force, swept along in the mania following the release of the movie, Top Gun. In high school, he entered the Marine Corps ROTC program and during two field trips, he had the opportunity to attend the 1st Marine Division Fire Day, shooting a M16 rifle and several other weapons. As he witnessed the power of the heavy weapons, his allegiance switched to the Marine Corps. His mother was not excited about his enlistment while his stepdad hope the Marines would mature him.

Eric had started drinking at parties in the 11th grade. The Marines are not adverse to kicking back with a six-pack of beer, telling stories and reliving memories. On leave at home, Eric enjoyed getting together with friends around a beach bonfire with a cold beer at hand.

Ask Eric about his three deployments to Iraq, his words tumble out - hard, hot, scary, boring, dangerous. When he chose to go into the infantry, he understood he was signing up for a job where he would be required to kill other men. He saw this as something to be done as part of his service to his country. He brought that experience home with him.

"The first deployment, I was just a happy-go-lucky kid. In that deployment, much of our combat was done from vehicles as we moved north toward Baghdad. The second deployment became much more serious. In the second deployment in Fallujah, we ran. The fighting happened as we did foot patrols and staked out positions on rooftops.

Fallujah was brutal as we were out on patrol nine days at a stretch with a short time off before we went out on patrol again. We were constantly operating at a heightened level of alertness. I was blown up several times."

At one point, on a rooftop, Eric believes that God spoke to him telling him to build his cover higher. He dropped into the house, scrounging for items to add to his defensive barricade. As he completed the task, a bullet zipped past his head. At eye level, this would have been a kill shot for the sniper had Eric not moved a second before.

Eric credits Iraq with making him grow up. When he returned home, post traumatic stress disturbed his rest, leaving him unable to sleep more than a couple of hours each night. He turned to alcohol to relieve the building anxiety.

" I had a lot of issues with how I was raised," Eric says. "I had some insecurity issues and a bunch of other things that I covered with alcohol. The drinking culture within the Marine Corps inflamed my anxiety over those issues. I began using alcohol to cope with the trauma. In turn, this caused more issues and exacerbating my post traumatic stress. On a scale of 1-10, my anxiety level was 25. I was miserable. I was angry, frequently blowing up at people, including my grandma! I had to quit drinking when I was forced to deal with legal issues. I came to the point where I wanted to quit because I saw what a terrible person it made me."

This is not unusual as anger builds out of combat trauma. Eric describes the Marines in Iraq as angry much of the time, always yelling about what was happening.

"The anxiety was 100% involuntary, a chemical imbalance in my body," he explained. "Lack of sleep contributed to the anxiety. At first, alcohol helped me to sleep but in time the alcohol increased the levels of cortisol in my body which interfered with my sleep."

As I listened, I recalled that the brain takes up to two years after a life-threatening injury or serious illness to heal. The alcohol Eric consumed extended his recovery. Unfortunately, many Marines returning from a combat zone find they struggle with alcohol. The alcohol helps muffle the pain and memories even as it debilitates their bodies and for some produces undesirable behavior. Given enough abuse, the Marine may find it difficult to stop drinking and enters a downward spiral. As Eric struggled with alcohol and his memories of the past, he began to recognize that his behavior was destructive.

"Over the seven years following discharge, I gradually became aware of my condition and the need to change. I seemed to constantly be in and out of jobs. I struggled with my relationship with my wife, Nicole. It seemed like we were on always on the brink of separating.

In late 2011, after their son, Bradley was born, Nicole left Eric, taking the baby with her. They were separated through much of 2012 and Eric slumped into a deep depression. He recalls feeling sorry for himself without actually accepting responsibility for the results of his behavior. The earlier anxiety ended but depression set in. He recalls, "There seemed to be no good days." Alcohol left him feeling a bit better. Eric and Nicole eventually resolved to continue living together but in 2016 he reached a new low.

Lost in self-pity, he took his pistol and slipped into the garage to sit in his car. As he listened to the music playing on the radio, he thought that ending his life might be better for everyone. As he pondered suicide, he thought of son, Bradley and his wife, Nicole, finding his body with his brains splattered across the inside of the car. His finger slipped from the trigger as memories from their life together flickered across his memory. He believes God brought the mental replay to mind. As the images faded, Nicole opened the door and began yelling for him.

"Not today," Eric thought.

Determined to seek relief, he began working with Bob Key, a counselor at the South Orange Vet Center and took his first steps toward sobriety. He didn't stop drinking but he slowed down. He learned that if he was struggling with PTSD, he should not drink at all. He began exploring therapy to counter the post traumatic stress, including group therapy and a procedure called eye movement desensitization and reprocessing. Through this therapy, the memories are moved to a different part of the brain where the veteran is able to recall past events without the trauma returning. Eric found that therapy began to help him recover.

His depression continued until 2021. That year, his grandmother recommended he apply for a week with a program through Samaritan's Purse in Alaska. The program brings veterans and their wives to the camp, expense free, where they spend a week in marriage classes learning to work as a couple. The program includes counseling and time for fishing, hiking and rock climbing. One day they flew into a National Park to watch the brown bears fishing for salmon. Camp personnel teach from the Christian faith, believing that God is the one who helps us work through the grief and problems in our relationships.

The experience was life-changing for Nicole and Eric. Did it resolve all their problems? No, but they returned with a better understanding of how to work together through the challenges in their relationship. Like many Marines, Eric has worked at rebuilding his life and goals. His credit rating has improved and he purchased a car that would carry the six members of his family. He found employment working for Amazon. He says the relationship with Nicole is awesome though they still have rocky moments.

Eric credits his work through the Veterans Administration programs, his AA sponsor and above all his relationship with God for helping him break out of the cycle of drinking and abusive behavior. As the reunion to commemorate their entry into Iraq approached, he anticipated reuniting with the men of the 1/5 as well as long time friends along the west coast. With this reunion, he remained sober, determined he would not slip with just one drink. An Alcoholics Anonymous meeting was included in his agenda for the weekend.

Chapter 3

The Fighting Fifth

One by one, we met men who had served in one of the four companies, Apache - Bravo - Charlie - Weapons. The Marines of the first three companies are riflemen, known as grunts. They are trained to march and fight. Weapons Company is concerned with heavy weapons, bringing in mortars and shoulder-launched rockets. In the field, a squad of men from one company may be assigned to support another company. Or, as in the battle for Fallujah during their second deployment, the companies may serve in widely separate locations, each with a mission to accomplish. On each of their three deployments in Iraq, 1/5 received some of the toughest assignments.

Ken Boles, Bravo Company, later explained that due to their training, the Fighting Fifth was considered an integrated combat battalion, operating seamlessly. The Fighting Fifth is the highest awarded Regiment in the Marine Corp and they are carry on this tradition with honor. As part of the Marine 5th and 6th Regiments, they wear the French Fouragerre, a braid over one shoulder, that was awarded for the Battalion's bravery during World War I in the battle of Belleau Wood.

This tradition means that the Marines of the Fifth Regiment may return to civilian life with damage to mind and body. They may carry the damage from combat for the remainder of their lives. Civilian reunions may be for sharing fond memories of good times in school and social organizations. The men of the Fifth carry the trauma of combat and the memories of friends lost. Their bodies still react to loud noises and intense flashing lights. They may wake in the night drenched in the darkness of a nightmare. Others develop medical conditions as a result of the environmental exposure

11

to toxic substances. Some are less afflicted than others. A number chose not to attend as the memories are too bitter to be revived.

In 2002, as Saddam Hussein remained President of Iraq, concern grew over whether his regime had manufactured chemical weapons and whether he would use them if caught in a conflict. The concern was not unjustified as Hussein had used biological and chemical weapons against the Kurdish people in northern Iraq. Ten years earlier, a coalition of forces from around the world had driven Iraqi forces out of Kuwait. At that time, they did not take the fight to Baghdad as political opinion weighed heavily against expansion of the conflict beyond Kuwait's borders. Now President George W. Bush and his administration believed that Hussein must be removed from power before another regional conflict threatened Iraq's neighbors.

In February, the United States and Great Britain began shifting forces into Kuwait as the politicians continued to negotiate, pushing Hussein to surrender his nuclear, biological and chemical weapons. On March 19, 2003, the time to negotiate ended. For forty days, the First Battalion of the Fifth Marines had waited in the desert as temperatures climbed, waiting for the command to GO!

On the evening of March 20, 2003, Lt. Colonel Fred Padilla met with the officers of the 1/5 to review the orders to cross the border into Iraq the next morning. A communication officer stepped toward Lt. Colonel Padilla with a message. Padilla waved him off, saying he would call the Command Center after his review. A second message arrived, again Padilla waved him off. Then the Comm Chief arrived and shoved a note at Padilla, demanding he read it NOW! A moment passed and Lieutenant Colonel Padilla turned to his officers, saying, "We go now!" An electrical charge swept through that tent as men straightened, glancing at each other and turning back to Padilla. Now!

Invasion / OIF 1, 2003

Three hours later, powerful engines rumbling in the dust, the first Amphibious Assault Vehicle moved forward carrying the Marines of 1/5 toward Baghdad. As Marine infantry units, beginning with Alpha 1/5, were the first to cross the border, the men strained to catch a glimpse of movement, of explosives as steel tracks crunched across the desert grit. In those first hours, 1/5 would experience the first casualty of the war as well as the heady sense of their firepower in destroying an Iraqi tank.

Their orders were to travel north, following a route roughly along the Tigris River while the United States Army along with a British contingent would travel along the Euphrates River, east of the Tigris toward Baghdad. Additional forces meant to enter Iraq from the north, traveling through the Kurdish enclaves but the Turkish government refused the United States permission to launch from Turkish soil.

The first objective for 1/5 were the southern oil fields of Iraq with orders to prevent the Republican Guard from destroying the wells and refineries. When the Marines first began to plan their assault, military commanders were uncertain what to expect in the resistance of Iraqi forces. The Marines of 1/5 were startled to encounter men emerging from trenches, their hands raised, some waving white flags. The early engagements were with conscripts that had not been well trained. Resistance was sporadic. The progress of 1/5 traveling north stunned those watching from home soil. Hope for the conquest began to build. Families across the United States began to hope that the conflict would end quickly and their sons could return.

As the Marines surged north along Route 6, the US Army struggled to achieve their objectives following the Euphrates through the region rife with Baath Party resistance At one point the Marines progress was halted and their forces were order to pull back to allow the US Army time to advance.

Lieutenant Colonel Padilla recalls that his heart was heavy as he ordered his men to retreat. No military force wishes to retreat, certainly not the US Marines! In their push north, they had by-passed areas of resistance. Now their delay allowed Hussein's forces to re-enforce Sadr City, east of central Baghdad. Muqtada al-Sadr, a Shia Muslim cleric, politician and militia leader, controlled a section of northeast Baghdad, along the route that would carry the Marines toward Hussein's headquarters.

Once they were given the green light to move forward, the 1/5 approached Baghdad and the resistance began to stiffen. They moved along the east bank of the Tigris River but Baghdad lay on the west side. Following Route 6, the Marines hoped to utilize bridges crossing what is known as the Saddam Canal and enter Baghdad. Following age-old military tradition, the Iraqi generals chose to make a stand at the wide waterway to protect Baghdad and began destroying the bridges.

As Corporal Chad Shevlin would later note, the Iraqis did not understand the capability of American amphibious assault vehicles nor the American military prowess in building bridges. Reaching the Canal, the AAV's splashed down the eastern bank and moved across the width of the waterway. In a sense this was ironic justice. Saddam had destroyed the huge wetlands supported by the Tigris River on which the Marsh Arabs, a tribal people, built floating dwellings that sustained their livelihood. As these clans did not support Saddam's regime, he took his revenge in re-routing the Tigris River to support other projects. This same waterway became one more defeat for Saddam.

As the Marines emerged from the canal they faced an on-slaught of Fedayeen fighters. The fight was fierce with bullets and RPG's slamming into the AAV's. They fought their way into Sadr City where the resistance became more volatile. Earlier, friendly Iraqis had run alongside the AAVS as they passed through villages along Route 6. Now the 1/5 was encountering what they had expected and the months of rigorous training paid off.

One incident occurred that seemed to validate the liberation of the Iraqi people as military leaders hoped. In his account of their push toward Baghdad in the book, *A Table in the Presence,* Lt. Carey Cash recounts a moment when a large crowd of Iraqis approached General Padilla's humvee.

"Lt. Col. Padilla and Sergeant Major Jones drove cautiously through the crowded city street, keeping tabs on where our companies were and checking out the area for the most secure place to set up a command center for the night. Just then a commotion broke out.

Padilla looked to his right and saw dozens of Iraqi civilians approaching his vehicle, frantically waving their arms back and forth up in the air as if something was desperately wrong. Sgt. Mayor Jones gripped his M-16 firmly in his hand and made sure the people could see its black barrel through the open window.

'Slow . . . slow,' Padilla instructed his driver. 'Let's see what they want.'

As the civilians came to the right side of the vehicle, they began to make signs and signals with their hands and arms.

'What are they doing?' Jones asked, well aware that neither he nor Padilla had a clue.

Padilla watched them, keeping the corner of his eye on his own pistol. It might not be as effective as Jones' rifle, but it could, if the moment went south, mean the difference between his life and theirs.

By now the crowd was more frantic than ever. They all seemed to be pointing wildly in the direction of a building at the end of dead-end road. Padilla searched the people's faces, trying to read them.

'Pree-zun, pree-zun, pree-zun!' Dozens in the crowd shouted the barely intelligible word while clasping their hands tightly around their wrists. They were indicating handcuffs or shackles or some other form of bondage.

'I think they're saying prison,' Jones exclaimed.

Several men, shouting the words, 'pree-zun', held out their hands at knee level, palms facing down. A sickening thought ran through Padilla's mind: Are they trying to tell us there is a prison full of children?

The more he watched the gesture, the more convinced he was that this wasn't a wild mob. These were parents, pointing in the direction of a building where they believed their children were being held. Immediately, Padilla called the unconfirmed report in on the battalion's radio, requesting the Human Exploitation Intelligence Team, along with a security platoon of riflemen, to get to the scene ASAP. . . . .

Although Padilla felt like the frantic crowd was legitimate, he and Jones had one rifle between them. He wanted to be sure they didn't find themselves fighting prison guards or other enemy forces that might have holed up in the facility. Still between HUMET and a platoon of armed Marines, he had to believe they would all be safe enough. So he directed his driver to proceed slowly in the direction of the building.

Closer and closer he and Jones inched, hemmed in by the hysterical crowd, toward the iron gates that marked the compound's entrance. The edifice definitely looked prison-like in its austere construction, with barred doorways and windows, raised walls, and stone walkways. Already a few of the men and women had begun to grasp and shake the gates with both hands, peering down a darkened hallway, shouting and raving at the top of their lungs.

Suddenly there was the sound of screaming-not the screaming of adults or women, but of children. Then, like a river bursting through a weakened band, as the iron gates swung open a mass of frail children came pouring out onto the stone terrace and into the arms of their waiting parents. They were malnourished and filthy, many of them wearing nothing more than rags of clothing. There were dozens at first; and then there were hundreds, mostly males between the ages of eight and fifteen. They flooded the lot in front of the building and soon were grabbing the necks of Padilla and Jones, crying, kissing them on each cheek, then jumping back into the arms of mothers, fathers, sisters, and brothers.

Padilla didn't know what to think or what to do. He couldn't help but feel emotionally overwhelmed. Little boys, many of whom were no older than seven or eight, were sobbing and clinging to their parents. He had never seen anything quite like it. Meanwhile Jones, in classic Sergeant Major fashion, bristled at all the overt affection. 'Come on now, boy, don't go kissing my neck. All right, all right, that's enough. Go hug on your daddy and mama.'

Padilla couldn't help but laugh, and neither could Jones; but their laughter was mixed with tears. Indeed, whatever elation the two felt as they watched the celebration continue was soon overshadowed by a grim reality. In the hours after the release, HUMET specialists found out the prison had been established years ago by the local Baath Party as a jail for the sons of parents who refused to support the regime. Many of the boys had been incarcerated for more than five years. As soon as they turned fifteen or sixteen, most of them were involuntarily conscripted into the regular Iraqi Army to man border posts or serve as cannon fodder for Saddam's wars."*

The former incarcerated were the conscripts that had surrendered to the Marines as they made their ways across the Iraqi oil fields. Think of this! My heart and mind rejoice that the United States could play a role in the liberation from tyranny. However, there were miles to cover and tough battles ahead.

Approaching Sadr City, 1/5 encountered steep resistance. On April 9, General Padilla received the order to move deeper into Baghdad with the objective of taking one of Hussein's palaces. He ordered non-infantry personnel to stay back within an area they had previously cleared. In the next few hours, the 1/5 ran a gauntlet of live fire as a hail of bullets and rocket-propelled grenades struck their transports as they pushed toward the palace. In that moment, the intense training they had completed at Camp Pendleton kicked in with Marines firing back at every blast even when the combatants were unseen. Calls for corpsman sounded over the radio as the wounded fought through blood and pain. At one point, Alpha Company took a turn that led them back along their route, snarling their advance. Undeterred, their objective remained the palace. Once behind the walls of the palace, the assault continued but now they found shelter and a means to evacuate their wounded as the men grieved the loss of Gunny Jeffrey Bohr.

As their tour of duty was completed, the men of 1/5 recognized that the fight was not done. The sense of jubilation would not last among the Iraqis. The most intense region of conflict centered around the towns in a triangle-shaped region dominated by the Sunni Muslims.

Six months after they returned to the United States, the Marines of the 1/5 were once again bound for Iraq to join the fight for the city of Fallujah. Located east of Baghdad, the city of 127,000 residents was a spiritual center of the Sunni sect with several key mosques. The men of 1/5 might have expected to quickly dominate the insurgency that sent residents fleeing but the conflict had changed since their entry into Iraq.

* A Table in the Presence, Carey H.Cash, Thomas Nelson Inc, 2004, P. 143-146

Fallujah / OIF 2, 2004

The invasion of Iraq in 2003 had been against Hussein's Army. When the 1/5 returned to Iraq in 2004 for a second deployment, the crowds of cheering Iraqis had disappeared and in their place came insurgency hidden within the local population. US Forces now faced experienced foreign fighters supported by local extremists. Marine Battalions 1/5, 2/5 and 3/4 were given orders to end the resistance in Fallujah, a Baathist stronghold.

Consider a map of the districts of Fallujah with four quadrants, beginning with the Jolan district in the northwest corner. Moving clockwise, east of Jolan, lay the quadrant known as Manhattan. South of Manhattan was the Industrial Sector. Queens lay south of Jolan in the southwest quadrant. Highway 10 ran east to west, intersecting the city. The Second Battalion was assigned to Jolan from the northeast, while 3/4 would enter Manhattan from the east side. The 2/5 located west of the city, would move north into the Jolan to meet up with 2/1. The 1/5 was assigned to the Industrial Sector, a warren of alleys, workshops and factories, all to be searched and cleared. A key part of the offensive was the cordon placed around the city to keep insurgents in without reinforcement from enemy forces. The population of Fallujah was warned to evacuate the city.

The Marines waited for the command to move against the insurgents. Each day, patrols encountered fire fights from insurgents tucked into homes and workshops. As they moved through the city, they never knew when a firefight would break loose. Casualties mounted and the politicians bickered.

This was no longer the conflict that our forces traditionally encountered due to three weapons with a deadly impact. In previous conflicts, the lines had been fairly distinct with each side seeking to advance and claim territory. With Hussein no longer in power, the insurgents had resorted to using guerrilla tactics.

As mentioned previously, Fallujah was a spiritual center for the Sunni Muslims. Each morning the call to prayer rang from hundreds of minarets. As the Islamic faithful responded, crowding into the mosques, the clerics proclaimed this a holy war, urging the population to take up arms and drive the infidels out of Iraq. Comparing their fiery sermons, in World War II, German citizens were led to believe that domination was their God-given destiny an an Aryan people. In the Civil War, preachers on both sides of the conflict had railed either against slavery or taught that slavery was God-given. What Coalition troops experienced in Iraq was far beyond the religious influence of earlier conflicts. The cries of the clerics were fanatical, their tirades driving the faithful to take up arms, believing that if they were killed, a reward await them. Negotiators struggled to make headway against these belief.

A second weapon was the civilian population of Iraq. Western forces believed that the civilian population should be kept apart from hostilities. The insurgents used the homes and people of Iraq as shields, hiding until they emerged to fire on the troops before ducking back into the alleys and buildings of a crowded urban center. With the insurgents, there was no attempt to spare the population of Fallujah. As residents began to flee the city, the men of 1/5 could never be sure which homes held civilians or insurgents.

A final weapon was the presence of Al-Jazeera, a Quatari-based media that favored the insurgency, airing videos of Coalition forces violently assaulting Iraqis while ignoring the abuse and terror inflicted by the insurgents. Late in 2004, the interim Iraqi government ejected Al-Jazeera from Iraq but the damage had been done.

With the three non-combat weapons favored by the insurgents along with the use of the IEDs, Coalition forces struggled to make headway. In turn, they were restrained by President Bush and his advisors. At one point, Commanders under General Mattis began to wonder if President Bush was receiving accurate information regarding the ground war from Ambassador Bremer and General Abizaid who both remained in Baghdad but did not experience the daily assaults on the troops. Ambassador Bremer was given authority to rebuild Iraq while General Abizaid was given command of the Coalition Forces. Bing West, in his account No True Glory, seems to indicate that neither man was completely aware of what the other was doing. Vast sums were spent with little progress in creating a functional government and rebuilding the damage to the infrastructure.

When Lt. Colonel Bryne received the order to move forward,

the 1/5 entered the Industrial district in the southeast, rapidly encountering resistance. One of their primary targets was a soda factory. From inside the building, the insurgents could fire on the 1/5. A shoulder-mounted missile blew the door off. The Marines surged into the building, their weapons swinging back and forth, seeking insurgents, ready to fire. From the soda factory, the 1/5 moved on to a mosque as a stream of men had been seen entering and leaving the premises. As they advanced, the Marines chose positions in homes and factories, using whatever they could find to build ramparts against sniper fire from neighboring buildings.

I am struck by one of the photos from OIF 2. The image shows several of the Marines scrambling through the detritus of a destroyed building, smoke and dust rising around them. One Marine, Ryan Ackerman, thrusts a hand into the air, screaming at his comrades. He seems to be pointing to fire coming from a nearby building. The reality of battle is chaos, fighters on both sides scrambling for position, men screaming, the boom of artillery, dust clogging the air and the lungs of the men in the fight. Chaos with a purpose, depending on the shared objective of each side to destroy the enemy before they are killed. I cannot fully grasp the terror of a man who tumbles to the ground as he feels the rush of a bullet passing a millimeter from his head, knowing a brief movement was all that separated him from oblivion. Every day is a scramble, the noise level pounding the brain, as cortisol surges through the veins of men seeking desperately to survive.

In April, Coalition Forces were ordered to stop their advance as negotiators were working through an agreement to expel the insurgents and re-establish the city government. Marine Battalions 3/4, 2/1 and 2/2 held the siege along the north and eastern side of the city. As they stood watch, one day melted into the next. Boredom and deprivation broken only by the terror of firefights as insurgents probed for weakness in their line.

With the temporary halt, the Marines were to patrol the city streets but this quickly became unviable due to daily attacks from the insurgents. Both sides positioned snipers, huddled in obscure positions, awaiting the moment a targets moved into the open. The Coalition forces might intend to stand down, the insurgents had no such intention. In the negotiations, several Iraqis, one a business-man, another a former Iraqi officer, stepped forward to offer their leadership. Neither was successful.

Wade Spence, Weapons Company, recalls how his attitude changed in Fallujah.

"Before Fallujah, 1/5 was sent to Okinawa to build up the forc-es there. Our mission was "winning hearts and minds." This was more theoretical than the invasion of Iraq during our first deploy-ment. During the invasion of Iraq, I was sympathetic to the plight of the people we met. They didn't choose to be in this situation. They would have preferred to live, working at a job, enjoying their families, going to the market.

"In Fallujah, we went out on patrol, took a position and then entered homes on inspection. I was assigned to an 81 mm mortar team. We were being cross trained for special operations, including the invasion of Iraq but also in riot control. Unlike the infantry, we were fighting from vehicles.

"Going in, we thought the fight was already won and were disappointed that we would not participate. Once inside Fallujah that all changed. It was game-on. Fallujah taught me to be thankful for our leadership under Lt. Glover. He was an awesome platoon commander. Our 81 mm team had a lot of flexibility. We became a quick reaction force, able to respond quickly to the violence.

"I became really pissed off in Fallujah, The politicians came in. They kept us from taking the city. The Iraqis did not take control of their city which would have been better for all. Another battalion followed us. When we left, we had paid the price, we knew the next battalion would pay that price again. I came home very bitter."

If the American advisors had been familiar with the history of Gertrude Bell at the turn of the 20th century, they might have understood that the foundation for the modern Iraq stood on faulty ground. Bell traveled extensively through the territory that would become Iraq and had a unique understanding of the tribes and sheikdoms that dominated the region. She spent years negoti-ating on behalf of the British government to pull stubborn sheiks into line even as the Wahhabi sect of the Sunni Muslims destroyed their competitors, assuming domination of the region. Iraq might have appeared to be a unified nation to a western observer but under the diplomatic facade, old rivalries remained and these formed the breeding ground for the insurgency that was tearing Iraq apart.

By November 2004, the daily assaults on the Marines had reached a point where Battalion Commanders could no longer countenance the official position of Ambassador Bremer and General Abizaid. The number of insurgents holding western forces at bay in Fallujah had doubled from 500 insurgents to 1000 with volunteers increasing that number to 2000 men. Lt. General Sanchez had seen enough and began to plan a second campaign in Fallujah, advising the Bush Administration that he would no longer stand by while Marines died.

The battalions, including the 1/5, that had first formed a cordon around Fallujah had returned to western shores. The second assault would enter the city from the north, expecting the insurgents to be unprepared for an assault from the north side of the city as it had been not been part of the approach six months earlier.

Five days after western forces began the revived assault, the Iraqi government declared the city of Fallujah under control. The fighting had been intense. Back on US soil, the men of the 1/5 held a degree of resentment as they considered the sacrifice, the hell they had survived only to be held back from completing the mission for which they trained. Due to the failure of months of negotiation, Marines had fought and died. For what? Fallujah and Ramadi would earn the reputation of being the most intense area of conflict during the war with the insurgents in Iraq.

The families of the men of the 1/5, the mothers and wives breathed a sigh of relief as they hugged their Marines. Traditionally, a Marine would only be deployed twice during his contract with the United States government. With the conflict in Iraq, expectations had to change. Five months later, as the 1/5 prepared to move to Okinawa, men called home to say that they had been redirected. They would return to Iraq with OIF 3, to fight in the town of Ramadi, west of Fallujah. More than one mother cried out in confusion and fear. Their sons had survived two deployments and now they were ordered to return to Iraq!

Ramadi / OIF 3, 2005

The 1/5 had led the invasion in OIF 1. They had fought to subdue insurgents in the streets of Fallujah. Now, in deploying to Ramadi, the conflict changed again. General Eric Smith, the commander of the 1/5 implemented a three-block strategy for Ramadi. One squad held a block secure as a second squad advanced into the next block with the third squad leaped forward to the third block. In turn, the first squad would leap over the third even as the second squad secured their advance.

The Marines are trained to fight and subdue their opponents, clearing the field of battle. They are not trained as a police force. Yet, this seemed to be their assignment in Ramadi. Employing the three-block strategy, they were to patrol the city, and put down resistance. Whereas the residents of Fallujah had been ordered to evacuate, in Ramadi, the population was left in place.

As Lieutenant Colonel Thompson noted, on one block Mardi Gras could be happening and one block over, we would be fighting for our lives. The Marines were now up against the most fanatical, experienced fighters and the fighting was intense. The men of 1/5 noted that the insurgents were changing their tactics from what the Marines had encountered during the initial invasion. In OIF 1, the resistance came from the Iraqi military reinforced by Fayadeen and jihad fighters. In Fallujah, the Marines encountered weapon fire and shoulder launched RPG's but in Ramadi their tactics changed as they began using more improvised explosive devices (IED). These explosive devices could be planted in advance and triggered as US Forces rolled past. In reading the autobiography of Maria Toorpakai, a Pakistani woman, I found a description of the impact of an explosive device.

"Depending on the point of impact, the bomb does its job in several ways. The initial blast wave: highly compressed air particles go out from the source at rates faster than the speed of sound, causing structural damage to whatever lies in its path–human or material. A series of supersonic stress waves follow. We haven't found a way to protect people from stress waves. These carry more energy than sound waves and pass invisibly right through the body, tearing up tissues and organs. After that, we see fragmentation from material packed in the bomb–ball bearings, nails, and razors– that travel at high velocity. Makes a machine gun look like a sling shot. Then you have secondary fragmentation from the buildings themselves–glass, concrete, metal. There's fire, of course, smoke and intense heat, which kills those trapped inside the wreckage. Finally the blast wind, a great vacuum of smoke and debris that sucks the polluted atmosphere right back into the initial explosion, leaving the any victims lacking oxygen."

Day by dad these Marines entered the city, drawing fire, look- ing for anything out of the usual. Consider one firefight described by Josh Shores from Alpha Company.

"The raid was to take place on a busy street named Cinema, in the heart of the souk. We were searching for IED-making mate- rials in that area after the company received a tip. Cinema boasted one of the tallest buildings in our area, around eight-stories high, and was also lined with many shops and restaurants. Large gran- ite tiles covered the sidewalk and made for an upscale-looking shopping area. Our platoon was to take the west side of the street, 2nd Platoon the east side. We rehearsed the movement up to the objective back on Hurricane Point, a rehearsal that was generally for the drivers to feel comfortable with vehicle placement, but all of us took part in it. We departed for the objective area around 1500, with the news of our fallen brothers still fresh in our minds and weighing on our hearts.

Quietly seated, we reflected on the day's events as the sev- en-ton truck chugged along down Route Michigan. The loud en- gine and smell of diesel made for a hypnotic and somehow relaxing break in the moment's commotion. Many of us smoked, so we typ- ically tried to get a few cigarettes in us while we rode in the back of the truck. Occasionally, a team leader would peer over the side of the seven-ton's walls to see what the streets looked like. If the streets were empty, we could expect a firefight or IED. If they were busy, that usually meant there would not be any enemy contact.

We pulled up to the objective and were given the order to dismount. We did this as quickly as possible because of our vul- nerability to an ambush while mounting and dismounting. Once we had both feet on the ground, we'd run in different directions to keep from being easy targets. Although we had received fire from a sniper a few times while mounting or dismounting over the course of the deployment, we were fortunate this raid was not one of them.

Once my boots were on the ground, I began my run toward the sidewalks while searching for cover. The street was not completely empty, with several vehicles lining both sides. I felt uneasy about everything and made the decision to charge my M249 SAW (Squad Automatic Weapon) into *weapon condition one* by pulling the charging handle to the rear and sending it back forward, making it ready to fire with the flip of my safety. We SAW gunners usually patrolled with our weapons in condition three, which just involved laying the belt of ammunition on the feed tray of the weapon and closing the feed-tray cover. The M249 SAW fires from an open bolt position, so pulling the charging handle back locks the bolt to the rear and makes it ready to fire. When you pull the trigger, the bolt slams forward and fires a round, then returns to the open position. This makes for a potential safety hazard if the operator of the weapon is complacent, like the Marine earlier in the deployment who had the negligent discharge. I would only put the weapon in condition one when we had contact with the enemy or it seemed enemy contact was imminent. I learned to trust my gut in Iraq; my gut was telling me to prepare for a fight that evening.

We all successfully dismounted while the drivers positioned the vehicles into blocking positions (a machine gunner and a driver always remained with each vehicle in order for quick movement and security of the vehicle). Our team assembled inside one of the shops, a photo-development business. I was with Sergeant Laws, two corporals (one of them a combat engineer attachment), and Lance Corporal Tager. We entered the shop and conducted a hasty search of the building to ensure there were no immediate threats. The upstairs was full of miscellaneous photo supplies and looked very disorganized. Two Iraqi civilians, most likely the shop owner and his boy, were present. They greeted us with a wave, saying "Hello" and forcing a smile. We were moved back toward the front door when Sergeant Laws heard through his radio that we were going to move south to meet up with the rest of the squad, who were in another shop. Because of short supply, only the Marines in the leadership billets were given a personal radio so we juniors always had to ask what was being communicated back and forth.

We formed a single-file line at the door, one of our platoon's corporals first, Tager second, the engineer corporal third, me, and then Sergeant Laws. I could faintly hear some communication over Laws's headset. The rest of the squad was ready for us to move to them. The first in line took off to the south and the rest of us followed. I had just left the shop and turned to face south to follow the combat engineer as he moved down the sidewalk when the insurgents' bullets began tearing into the vehicles directly to my left.

A heavy amount of automatic gunfire poured into the street, ripping into the parked cars to my left and the buildings to my right. It felt like I was in the middle of a tunnel, being riddled with gunfire. I was just leaving the building, so I turned back toward the photo shop and took cover behind the service desk inside, alongside Sergeant Laws. The gunfire sounded like a torrential downpour on a tin roof, but amplified to a deafening roar. The snaps of rounds passing by the shop were accompanied by whizzes from ricochets caused by impacts with vehicles, ground, and buildings. Rocket-propelled grenades (RPG) were being fired by the insurgents and were impacting the surrounding buildings. It felt as though an hour had passed before Sergeant Laws and I looked out the storefront's large windows. Gripped by the intensity and destruction unfolding before us, we only sat for about ten seconds before bullets began impacting the windows of the shop. Unsure of where the enemy was located, Sergeant Laws and I both fired through the windows at the buildings across the streets. It seemed the most likely spot because of where the rounds seemed to impact the windows. My ears rang immediately after my first burst with the SAW. We didn't have any hearing protection, so the first burst of gunfire always hurt the most, especially if firing from inside a building. I let a couple more bursts go and realized I wasn't really shooting at anything; I was just scared and I knew I needed to conserve my ammunition. We also realized 2nd Platoon was most likely across the street and possibly on the second floor.

Sergeant Laws could not get radio contact with anyone else.

Neither of us had a clue which building the gunfire was coming
from. He decided we should try to get to a different position within
the building, so we made our way up the stairs to the storage area.
Before I left the comfort of the service desk, I remembered the two
Iraqis in the shop and wondered where they had gone. I noticed
a black curtain behind the service desk. Only then did I realize
someone could have easily killed us by us not keeping watch on
them. I figured the shop owner and kid were behind the curtain.
To have more control if I needed to fire from the hip, I positioned
my weapon under my arm and reached out with my left hand to
pull the curtain open. With my guard up, I was prepared to elimi-
nate any potential enemy that could be hiding behind the curtain.
I threw it to the left and immediately returned my left hand to the
handgrip on the weapon. I was breathing heavily, my eyes were
wide to take everything in, and I was standing tall before the two
Iraqi civilians who now sat on the floor gripping each other in fear.
The older man had the younger boy in his arms, in a very protective
and vulnerable position. The man looked up at me.

"Mister, no," he said softly, extreme fear etched on his face and
reflected in his voice. I stood there and processed everything, being
sure not to miss a threat.

"Mister, no!" he said more confidently, fearfully begging for his
life, with my weapon still aimed at both him and the young boy.

"Stay!" I said and gestured with my left hand to keep down.
The man nodded and looked back at the boy. I made my way up
the small flight of stairs to where Sergeant Laws was. The volley of
gunfire was still present outside, and now I could hear our heavy
vehicle-mounted machine guns opening fire. Sergeant Laws still
could not get hold of anyone on the radio.

"What took you so long?" he asked. I told him about the two
people downstairs.

"Damn, dog, I forgot about them. Good thing you didn't," he
said as we both smiled.

"We have to go find the other guys," he said, knowing we'd
have to move back onto the street and meet up with the other
Marines from 3rd Squad.

"Okay," I said, and we made our way back down the stairs. I

looked back, waved goodbye to the shop owner and made my way to the door where Sergeant Laws was already standing.

"When there is another lull in the gunfire, we are going to run. Ready?" he said.

In my mind, I doubted our survival, but still replied yes. I didn't think about home or my fiancée or my childhood in those moments. The image of being ripped to shreds by bullets or experiencing a slow and painful death out on the street alone came to mind. I was the most scared I had ever been in my life. The bullets were still impacting the cars and street just feet from where we stood. I felt sick, most likely from the dump of adrenaline catching up to me. I wished our communications had worked so we could figure out a better plan. Instead of just Sergeant Laws, I had hoped to be with more Marines to increase our chances of survival in any situation that awaited us. I started wondering if my gear was too heavy to run fast. My boots felt heavy. Everything felt heavy. That's when the incoming fire momentarily stopped.

Sergeant Laws took off running, and I was just behind him. I was running out the door and committed to whatever lay in wait for us. My eyes darted around to locate any enemy positions, and places I could use for cover. Tunnel vision. My view was obstructed by a lot of smoke from exploding RPGs and debris from gunfire. In an attempt to make myself a more difficult target, I started running in a small zig-zag pattern. It was surprising how slippery the tile sidewalk had become, with all the debris from the buildings scattered across it and with the soles of my boots caked with tar from earlier in the day. I slipped and fell in the open.

My world began to operate in slow motion. I watched as Sergeant Laws ran down the street.

"Wait!" I shouted, but my call fell on deaf ears. I watched as the building at the "T" intersection of Michigan and Cinema exploded with the impacts from our truck-mounted Mk-19 40-millimeter automatic grenade launcher, Captain Thompson's vehicle. Sergeant Laws's silhouette caught my eye as it darted into a building. I scrambled to get to my knees and get a footing again as the enemy gunfire picked up. As soon as I had my footing, I was moving to where I had watched him enter a shop. It felt like it took minutes to get there, but it was only seconds.

"Coming in! Shores coming in!" I screamed as I neared the door to the shop. I plowed through a stack of Marines who were waiting near the door.

"Jesus, where were you? I thought you were hit!" Sergeant Laws exclaimed.

"I fucking fell. Thanks for looking back, asshole!" I replied." *

This incident was not unique with firefights occurring almost daily. Squads were rotated in and out, given a couple of days off to rest away from the fatigue of battle. The three block strategy did not seem to subdue the foreign fighters hidden among the civilian population or allow them to progress to a region free from conflict. More men from the 1/5 lost their lives, morale sagged.

As their deployment neared termination, each man counted the days till they would leave Iraq. Then, another IED triggered and five more Marines lost their lives. With this fresh in their minds, the men prepared to leave Iraq and return to Camp Pendleton and their families. A select number would return as private contractors but the majority prepared to resume civilian lives.

Twenty years passed and now the Marines of the 1/5 have returned to the company of men they fought alongside. At 0600 hours the next morning, the men would assemble at the foot of 1st Sgt's Hill, making the ascent as they had in their training years earlier.

* Counting on Death: A Marine Infantryman's Journey From the Front Lines of Combat to the Fight for Peace. Shores, J., Casemate. Havertown, PA., 2025

Chapter 4

Muster

As the Fighting Fifth returned to the United States after their third deployment, many would recall the camaraderie but each man also carried the trauma of combat and the memories of friends lost. In their time in Iraq, each Marine had grown in experience, far different from the boy who first enlisted. Now, their bodies reacted to loud noises and intense flashing lights. Some woke frequently to the darkness of a nightmare. Others developed medical conditions as a result of the environmental exposure to toxic substances. They had been trained to fight, using hands, side arms, long guns, heavy artillery and explosives. But there had been little training to counter the damage wrought by combat.

As a result, few were prepared for the psychological and neurological damage done by trauma, injuries and explosions. The challenges these men faced as they returned to civilian life after their service to their country were daunting. When a man is discharged from military service, he is expected to pick up where he left off before enlisting. He will find a job, either rent or buy a home and if unmarried, he may choose a partner. However, he may not be prepared to be successful in these tasks. Ryan Ackerman, Alpha Company, recalls the day he returned to his wife in Indiana.

"I enlisted in December in 2000 right before I turned 17 years of age. I was very serious about becoming a Marine and serving my country. When I was discharged I didn't know how to find a job, how to put together a resume. Who would want to hire a man trained to kill?"

When I consider this statement, I realize many Marines, men and women, enlist right out of high school. They have not accumulated job experience as a civilian. What do they put on a job application? Serving four years in the military, particularly in the Marines, ingrains qualities in a man that are attractive to an employer. The men learn to get up promptly and show up for duty. They understand the value of working as a team and the training to do the job. However, in terms of actual work experience, the grunts may struggle with listing previous employers, complete with contact information. There is a limited need for knocking down buildings and killing insurgents.

As part of preparation for life beyond the Marines, a good education is a foundation for the job market. When a Marine is deployed every six months, it is difficult to complete college classes toward a degree. If a Marine has set aside part of his income in the GI Education program, he may have the finances to attend either a trade school or community college. These funds are limited and may not completely cover living expenses.

Many of the Marines gravitate toward employment providing security for corporate and private clients. Some join a police or sheriff's department. The training and experience they gained in combat is supported by the hypervigilence that follows them home from Iraq. However, there is a dark side to working in private security as the stress of being hypervigilant may continue to support the trauma and stress they experienced in Iraq.

Some veterans use their knowledge of firearms to start a business or train others in using firearms. Michael Tager, Alpha Company, began dealing in firearms out of his home but as the business grew he was able to open a commercial location in 2024. He would learn the challenges of supply and demand in the civilian market and the less than ethical tactics exercised by the competition.

Some Marines, pursue careers such as fire fighting, with an element of risk. In the state of Washington, the Bureau of Land Management developed a team of veterans to fight wildland fires during fire season. Springing into action when a fire erupts appeals to men who once knew the sense of camaraderie in combat. They fight the destruction of our forest on our behalf, serving our country as they once did in military service.

The Veterans Administration has drawn up programs to aid in

acquiring a job. One employer I met was willing to hire veterans for his company that provides service and equipment to microwave towers. He was willing to train these men on the job but he found the reality of hiring vets was not what he expected. There was no clearing house to reach out to veterans and those that happened across his advertizing often did not meet his basic requirements. With a central clearing house to provide contacts, employers would be able to ask basic questions in determining whether the man would be one they would consider. Based on the track records of other government programs, I don't believe that government would provide the best answer for such expectation. A central clearing house would preferably be run by business-oriented individuals with a platform for performance accountability.

Another problem for veterans arises with managing finances. Most 17-year-old men and women have not received an education in managing their paycheck. The military provides some of the basic necessities like housing. When they are discharged, the veteran is responsible for organizing his life, including the demands for housing, food, and insurance. This understanding starts far earlier than the date of enlistment. A basic understanding of finance should be required in our public high schools before a student can graduate. Without that training, a veteran may find himself in debt, wondering why the money from his pay check does not stretch to cover the expenses.

As veterans struggle with finding a job and managing finances, the two can lead to a secondary problem with a man ending up homeless on the streets. When we pass a young man in a military camouflage jacket, standing on the corner of a busy street, holding a sign asking for a handout, we wonder why he doesn't get a job. A veteran may slip through the cracks, struggling with mental health issues or be simply unable to counter the challenges until homelessness leaves him without the permanent address that most potential employers expect.

Those returning to their families or in a committed relationship often experience a different kind of homelessness, not one of a physical dwelling. In returning to his home, the wife or partner may find the Marine altered from the man who left to serve in a war zone. The violence of combat, the images pressed into a man's memory make an indelible impression on the mind. The couple established a home with their roles in place. When reunited, so much has changed. The partner does not understand her husband's struggle because she did not share in his experience. Some women wish to be sympathetic and supportive while others decide not to remain in the relationship. This is often an even greater loss than a physical dwelling. The Marine has lost a sense of refuge in the one he loves.

Marcus Sudani met his wife after discharge from active duty. He knows combat effected him mentally. His wife knows the man he is today, post Iraq.

"I know what he is now. I did not know him before he went to Iraq," she says. I believe we had an easier time adjusting to how he has been effected because that is all I've has known.

I watch them joking back and forth about learning to blend together as a couple. They tease each other about how they handle the stress in their relationship. After hearing stories of heartbreak from Marines who struggled in their marriages and many divorces, I can relax and enjoy their easy banter.

The National Center for Biotechnology Information reports that 2.4 million military personnel have been deployed to Iraq and Afghanistan since September 11, 2001. Of those, 772,000 have sought medical care through the Veterans Administration. Men and women have suffered long term injuries from amputations, spinal cord injuries and traumatic bring injuries. Doctors are using innovative surgeries and equipment to treat many of these men and women.

While external injuries are evident to the human eye, the damage to the human brain after being exposed to repeated explosions can be more difficult to diagnose and treat. Many of the infantry men experienced traumatic brain injuries. Doctor may say that the TBI's heal in 60-90 days. Veterans know from experience that the long term effects can last a lifetime.

As our scientists develop complex chemical formulas used in the energy sector and in warfare, our veterans are often exposed to substances that cause physical harm long into the future. Dust storms caused many to develop respiratory complaints. Veterans from the Gulf War in 1991 complain of fatigue, headaches, joint pain, indigestion, insomnia, dizziness, respiratory disorders, and memory problems. The range of symptoms has been labeled the Gulf War Illness (GWI) and is believed to be a result of exposure to toxic chemicals. Now the Marines sent to Iraq and Afghanistan share many of the same symptoms. Doctors are working to understand the best treatments for GWI.

The Veterans Administration has developed criteria for neurological injuries. Veterans who develops neurological damage related to Parkinsons, Multiple Sclerosis and other illnesses within seven years are classified as developing combat-related disorders. Beyond seven years, the VA does not classify the illness as caused by time serving in a combat zone, yet every veteran seems to know another with neurological damage, leading to a chronic condition.

Much has been said and written about post traumatic stress (PTS). The indications of post traumatic stress are grouped into four categories by the Mayo Clinic: Intrusive memories, avoidance, negative changes in thinking and mood, and changes in physical and emotional reactions.

Intrusive memories may include the re-playing of events through the brain on an unending cycle. Flashbacks may leave the Marine re-living an event in real time. Nightmares may disturb his sleep cycle or he may experience emotional and physical distress due to a particular trigger such as a loud noise or an unexpected appearance. As a result, the veteran begins to avoid these potential triggers. Outward indications may include insomnia, irritability, over-reaction to outward stimuli, lack of trust, avoidance of people or intimate conversation, and abusing alcohol. Some find that the playback of images and incidents they have witnessed plays as an unending loop in their minds. I've seen men explode over the smallest trigger and witnessed those who sink into silence, avoiding personal contact.

Many of the men who returned from World War II did not talk about their experiences. This is the second indication, known as avoidance. They found employment in the post war industry booming across the United States in the 1950s. Their brains would not allow them to forget but they did not verbally express their struggle. Negative changes in their thinking and moods deepened, leaving their families puzzled as to why they were unwilling to interact and unwilling to tell their stories of combat.

Negative changes in thinking often included thinking badly of oneself or others. Some experience difficulty in recalling what they lived through in combat. Most veterans who experience these changes have trouble maintaining close relationships. This has been the cause of many broken marriages and estrangement from families. The veteran no longer seems to have interest in activities he once enjoyed. His family complains that he seems numb or lacking in empathy for what is happening around them.

Higher levels of cortisol run through many Marines. For some, the strain on the adrenal gland is overwhelming, leading to adrenal crash. Marine Victor Dedra began developing adrenal crashes. This serious medical condition is often misunderstood by medical practitioners.

He describes the condition saying, "I start by being a total asshole to my wife. I become very hungry, followed by low blood pressure and a fever. This leaves me exhausted." He chose to move to Costa Rica where the pace of life is quieter. Fewer triggers.

As part of their search for relief, many veterans indulge in destructive actions, drinking too much alcohol, participating in reckless behavior and make unwise decisions that place them at risk. Trouble sleeping is common, leading to irritability. Consequently the veterans struggles with guilt over their actions. They may not recognize the connection between hypervigilance and the destructive behavior so they do not seek assistance.

I began with the story of Eric Young, who served three deployments with 1/5 Alpha Company. He demonstrated a number of these reactions after he was discharged. He explained that in drinking, his body was releasing ever more cortisol, keeping him on edge. He realized that he had to stop drinking if he wanted to preserve his relationship with his family. He sought counseling and ultimately chose to commit to the Christian faith.

The veteran's reluctance to seek help is not uncommon. Many seem to believe that in seeking help they will show weakness.

When they do enter the Veteran Administration network, seeking an appointment, they find that the delay may last months before they are scheduled to see a provider.

Kenn Boles, 1/5 Bravo Company, descended into a cycle of destructive behavior after his discharge. The day came when he walked into a facility and announced, "I need to talk to someone NOW!"

With the increase in suicide and destructive behavior toward others, the staff of these facilities have begun to take immediate action. Boles was escorted into a provider's office. He tells other veterans, "if you need help, don't waste your time asking for an appointment. Walk in and tell them you need help immediately. Don't wait!"

Once the veteran registers for assistance, the Veterans Administration offers several programs and seems to favor what is called evidence-based treatment with the first phase consisting of three therapies. These are identified as cognitive behavorial processing (CBT), prolonged exposure and EMDR which stands for eye movement desensitization and reprocessing.

With cognitive processing, the veteran is asked to recall traumatic events in combat and then write a statement describing the event. In a following session, he reads the statement aloud and a period of critical thinking follows as the veteran is challenged to identify the thought process regarding the experience. It may be that the veteran blames himself for the event or he feels inadequate to address what transpired. Through therapy he comes to understand that he cannot be held responsible for everything that happened. He begins to release some of the emotional weight. CBT may be more effective when combined with stress and anger management therapy, biofeedback and group therapy. When CBT is less than effective, therapists may introduce present-centered therapy which is less confrontation that the methods used in CBT.

Prolonged exposure therapy (PET) encourages the veteran to re-tell the experience in the present tense, gradually examining what happened. The session may be recorded so that the vet can listen and process his memories with the hope that he will grow more comfortable addressing them. This is particularly important when a vet is avoiding the memory that is disrupting his subconscious state.

EMDR is often used along with the other two therapies. The veterans is given two small vibrating devices, holding one in each hand. As an alternative, the therapist may tap the patient's hands if the device is not available. The veteran is asked to relate an experience as the vibration alternates between the opposing sides. The science behind this shows the device helps the brain to transfer the painful memories from front of the brain to a part that can regard the incident without the emotional attachment.

Along this same approach is the Theta-Burst transcranial magnetic stimulation using an elecromagnetic coil that targets specific points on the skull to stimulate the brain with magnetic bursts. This helps to change the patterns in thinking of past experience and how a veteran views himself today.

For those in a state of hypervigilance, mantram therapy may help stabilize the patient's arousal. With mantram therapy, a patient is asked to repeatedly cite a phrase that is meaningful to him, allowing his mind to relax. Forms of mantram meditation are found in both the Far East and in the Orthodox tradition.

Medication is an option though many younger vets are reluctant to use the drugs for anxiety or sleep disturbances. The drugs do have some significant side effects for some individuals and finding the right medication can take time. The Veterans Administration has dispensed a range of antidepressants for depression and anti-anxiety drugs for those who remain either hypervigilent or are ridden with anxiety.

The discussion over psychedelic substances is becoming more mainstream in the last few years. There are four classes of drugs that produce hallucination. Briefly, ayahusca is a plant-based substance out of Latin America believed to help addiction, anxiety and depression. LSD, the popular drug among the flower people of San Francisco in the 1960s alters mood, perception and our conscious state. Psilocybin or specific mushrooms, similar to LSD, alter consciousness and how the user perceives their circumstances. Finally

MDMA is not technically a psychedelic but give the user a sense of euphoria, alters perceptions, and increases arousal and the desire to interact with others. Of these, ayahuasca is the least adapted to medical treatment.

Many practitioners have turned from psychedelic drugs to ketamine. This drug is not a true psychedelic but was developed as anesthesia for surgical patients, leaving them with fewer hallucinations. A fairly high dose will induce unconsciousness as happened with Matthew Perry. Practitioners advise that the drug only be used in a clinical setting with an attendant present. In a patient struggling to address the causes of depression, this drug opens the neuro pathways in the brain. Ketamine can lower the inhibitions toward talking over what is painful to remember.

Usually, a patient using ketamine will experience a high, lasting between 1-4 hours and may display a range of both mental and physical reactions. Some guides may talk with their patients helping them understand what they are experiencing. The effect of the ketamine lasts for two to three days. During this time, talk therapy is employed as the patient is encouraged to verbally express what is causing his struggle. Ketamine is also used in treating depression, substance and alcohol abuse, and anxiety - all conditions that may result after exposure to combat.

I respect the advice of one nurse with 18 years experience. She insists that it is important that one who receives ketamine should first make an effort to establish a healthy lifestyle before submitting to a mind-altering drug. She includes eating properly, getting enough sleep and not drinking as all these may alter the effects of ketamine.

As I learned about this drug, I thought how some are willing to open the neuro pathways and do the hard work without the assistance of a drug such as ketamine or a hallucinogen. Therapists explain that not everyone is able to do this. It is important to remember that ketemine should not be used when alone due to the possibility of a negative reaction or that the patient may temporarily lose consciousness. It should be used with talk therapy to gain the most benefit.

Some veterans are using marijuana to sooth their anxiety and many states have passed medical marijuana laws. I am not completely convinced that this therapy is right for most veterans as our bodies adapt and the herb may become less effective over time.

Canine therapy for those who enjoy the company of a dog is becoming widely used. In recent years many have begun to blur the the distinction between a service dog and an emotional support dog. Service dogs are highly trained for nine months, beginning with basic obedience. If a dog successfully passes this first phase, he is trained specifically to the needs of a veteran. Once the dog has completed training, he is officially registered as a service dog. The training is expensive, which mean the demand outpaces the supply.

Emotional support dogs should be trained to obedience, then paired with the special needs of a veteran. Unfortunately, we've seen the lines blurred on what is an emotional support animal. These are not dogs that are casually picked up from the local humane society and labeled as a service dog because their owner likes the dog.

Coco, a German short-haired pointer, has become very popular with the Marines attending the 1/5 reunions. He is the service dog for Wade Spann, Weapons Company. On an early morning at the base of 1st Sergeant's Hill, Coco is a willing recipient of attention. Spann describes Coco's role in his life.

"After I was diagnosed with MS, I qualified for a service dog. I knew I needed to stay active. I needed to be better at stress management. When I first met Coco, she was 14-months old with these big brown eyes. One look and I knew she was my dog. She had been assigned to another vet but his wife developed cancer and he felt that it was not the right time to add a service dog to their family. I trained with her for several months. She is 6 now, I've had her for 4-1/2 years.

"She travels everywhere with me. She keeps me in a routine. I have a short temper and when I'm upset, she will jump up and place her paws on me to distract the tension. She senses my mind set and works to soothe me if I'm upset. Just giving her a big bear hug helps to relieve the stress. She knows I'm damaged and she'll sit on my lap, keeping me from the breaking point.

United Airlines has been very good with Coco as my service dog. She meets a lot of people and responds well to other people when they are under stress."

During a hike up the hill, Coco appeared to be like any other dog, exploring new scents as she raced along the trail. However, despite her joy in the moment, the pointer never lost sight of Wade, checking back on him frequently. Unfortunately, the requests for a service dog outstrip the current supply as training is intense.

Post traumatic stress has been shown to be quite disruptive in 20 percent of our veterans from Iraq and Afghanistan. In a study, several factors contributed to whether a man developed PTSD, including serving in combat and participating in a traumatic event, being seriously injured, and knowing someone who had been killed. Being married and being the parent of small children seems to play a role as well. Four factors credited with decreasing the liability of PTSD included graduating from college, whether a man understood the mission in which he served, being an officer and a strong religious faith. The stories in the following chapters may question these last four factors. Regardless of each man's unique outlook, therapies seem to have improved the lives of many Marines allowing them to make the transition to civilian life more easily.

One injury that has not received much attention rises out of the concept of guilt. We are given a conscience when we are born. This enables us to know the difference between right and wrong. Our conscience understands that killing another human being transgresses the sacredness of life. Most Marines will tell you that they killed because they knew that their adversary was intent on killing them. This does not address the deeper issue, the taking of another life.

In studying the effects of religion and spirituality on health and our human psyche, Psychiatrist Harold Koenig writes, "The anguish, the moral injury that people experience, in some respects, that's a good thing. Because part of the repentance involves having some level of conscience that says, 'I've done something wrong. I should feel guilty."

If we fail to consciously acknowledge and address this moral injury, our psyche begins to act out through mental illness, through violence and through destructive patterns of behavior. Some would scoff at this idea but with a renewed sense of spirituality flowing through our society, we could examine what is necessary to heal a moral injury. In addressing this concept, we are able to confront the problem of evil and seek absolution.

I asked Corpsman Al Alcala about his biggest challenge in adjusting to civilian life. His answer is echoed by other Marines.

"I find such a lack of respect. I've served my country. Strangers don't know what I've accomplished and they treat me like I don't know anything." He speaks with a heavy Hispanic accent.

A lack of respect runs rampant through our society in sharp contrast to the discipline of the military. If a Marine fails to address a superior properly or salute an officer, a man can face disciplinary action. To be called stupid when you once oversaw a crew of corpsman assigned to a battalion is a sharp blow to a Corpsman's pride in what he has accomplished. There is no explaining the training and experience within the Marine Corps to a civilian so that they will fail to fully appreciate what lies within the brain of a combat veteran.

Ultimately, not every Marine from the 1/5 choose to attend the reunions. Considering the challenges many have faced after discharge, this is not a surprise. A man might be concerned that seeing old comrades will either be too painful or bring about further disruption in to his life. Those that did attend often found they had returned to a comfortable place where they were understood and accepted.

Every Marine has a story to tell. The stories have many details in common but each man's effort to find his place in civilian life is his own, different from the men he fought alongside.

Chapter 5

Undaunted

Ryan Ackerman gave his mother little choice! He was going to enlist in the US Marine Corps at age 17, regardless of her concern for his safety.

Ryan was born and raised in Indiana. When he was two years of age, his parents divorced. After his mother remarried, he remembers mother and stepfather as very strict, abiding by cultural expectations. Ryan later realized that he had shown signs of attention-deficit disorder and that his parents must have had their hands full in raising him along with his sister. At age 16, he was certain he wanted to enlist in the Marines and saw no reason to wait. His mother thought differently, insisting that he finish high school. In December 2001, he headed for boot camp.

Four months later, two planes flew into the Twin Towers in New York City. That day had been peaceful with the recruits going through the physical and educational sessions. Late in the afternoon, a drill instructor (DI) asked the men from New York to step forward and led them into his office, delivering news of the attack privately. After they returned, the DI demanded, "Where are my infantry guys?"

Ryan, along with several others, stepped forward. "Get ready. You're going to war," the drill instructor told them. He then explained what had happened early that day as foreign national hijacked four planes, crashing into the Twin Towers and Pentagon.

In March 2003, as the amphibious assault vehicles crossed the Iraqi border, he remembers the silence of the Marines in his transport. The air filled with diesel fumes as aircraft thundered overhead. In this surreal moment, Ryan felt as if he was out of himself. The Marines understood the world was watching, their movements scrutinized by millions of eyes as if in an ancient arena. Their deployment was characterized by movement, pushing ever forward. The first sounds of gunfire erupted after crossing the border as the men of 1/5 engaged enemy combatants.

He deeply admires the Battalion's former commander, saying, "By far, Fred Padilla was the most impressive man I had ever met, a great role model. He made the lives of the Marines under his command better with the way he lived. I would do anything for him. He was a great role model and it was such an honor to work under him."

A year after the initial invasion, the 1/5 returned to Fallujah. Ryan remembers that with the deployment, the adversary seemed to be change from local Baathist to insurgent to mercenary. In Fallujah, his deployment, Ryan was assigned to Lt. Philip Treglia, the commanding officer for 1/5 Alpha. Captain Treglia responded to nearly every incident in which Alpha was engaged, his squad of men either driving fast or racing on foot through the streets.

"Every time we had contact, we took off running. It was just our squad and it was wild. If we came into contact, we had to find a way through. When you're in a fight, everyone is expected to do their best. You're expected to step up."

The action remains a blur as one day flowed into another. The men were at risk from both ambush as well as spontaneous eruptions of violence as they passed through the streets. In relating those events, he reveals the exhaustion of constantly responding to violent attacks even as the action made him feel alive in daily confronting death.

"Iraq changed me," Ackerman recalls. "Changed me in many different ways. I'm still learning how it changed me."

After his term of service ended in August 2005, he did not re-enlist. Leaving active duty, he realized he had no experience in filling out a job application or creating a resume. He had enlisted with the delayed entry program while in high school. His job experience was non-existent.

"How do I take my experience in the Marines and put it in a resume?" he asks, thinking back to that time. "You have to take those good qualities you learned as a grunt and restructure them to fit a civilian job."

Returning home to Indiana, he thought he might like to become a US Marshall as his experience with the 1/5 would be good background for a career in law enforcement. As he checked into what was required, he realized it would be a lot of work to qualify.

"I didn't have the confidence in myself to do it alone. I felt very vulnerable," he says.

Instead, he went to work in a friend's tire shop. A mentor suggested he apply to Toyota for a position in their paint department. He interviewed for the position, was hired and moved south to Arlington, Texas. He remained there for over four years. His wife did not like Texas!

General Motors had entered a partnership with Toyota and their financial vulnerability soon dragged the company into bankruptcy court. Leaving Toyota, he interviewed with a private security firm that provided executive bodyguards for high-profile clientele. He passed through their Academy and qualified as a bodyguard, working for some high-profile clients.

Unfortunately, he did not like what he saw as some of his clients seemed to regard themselves as entitled to privilege even as they treated their staff badly. He felt as if he were a puppet. For a Marine with combat experience, this was not acceptable. He resigned.

He worked with a start-up company building electrical vehicles but this company also went bankrupt. Another job offer came along, this time from a start-up called Tesla. For four years, he worked as part of a team, build Tesla's early version of the Beta model and self-driving vehicles. His team did behind-the-wheel testing on the vehicles before each was delivered to clients.

Working his way up to Quality Control Manager, he helped to develop the test facility in the Netherlands and the team that launched the S model and the X model. However, in working with Tesla, he began to realize that he had no life other than work. He had little time for family and his daughter would not wait to grow older. Once again, he resigned, seeking a better balance between work, family and his other interests.

"I started listening to my demons," he says. I began sinking into a depression and asking why I should keep going."

His thoughts echo those of other Marines who struggle with employment and relationships after leaving active service. In the darkness, the phone rang. His former commander, asked, "How you doing?"

Philip Treglia could not be ignored. He dived into Ryan's struggle, seeking to help him regain his sense of purpose.

"You can have friends but how many check up on you?" asks Ryan. "Hearing from Lt. Treglia - that's love in action. The call woke me up. In the years since leaving the Marines Corps, I had pushed the military out of my life, intent on making money. That phone call helped me think about what was important. I wanted to mobilize the Marines with whom I had served, to serve the men of the 1/5, both past and present."

In 2022, he called General Padilla and brain stormed about forming a non-profit 501C where the Marines of the 1/5 would find a safe place to talk and exchange memories, life experience and times when they need a bit of help. Both men believed there needed to be a safe place to meet. They wanted to provide internal resources for the alumni and purchased a domain, setting up the 501C.

Gunny Homer Valdez wanted shirts made for his men, featuring a logo for what is now known as Apache Company. Ryan worked with a tatoo artist to come up with the logo. The tattoo artist created a design showing a skull with a feathered headdress and a combat knife. He set up an awning at one training session and sold the shirts as he chatted with the men about their experience. This led to other products, now featuring the work of Ishmael Enevides, a Marine from OIF 3, who provides artwork for Warface.com and the 1/5.

Even as he dived into the new venture, Ryan found happiness with a new partner. With the success of the 2023 reunion, the 1/5 Association Board began working toward the next reunion, commemorating their time in Fallujah, Iraq. Today, Ryan flashes a bright smile as his mind clicks over all that needs to be accomplished.

He see the confidence he has acquired since he first left for active duty. He makes a point to start each day practicing self-care, small acts of kindness and building long-lasting relationships. Ryan credits the incredible love, unwavering strength, and boundless support his family has provided, calling his beautiful wife his closest confidante and his rock. Their children and an Australian Shepherd puppy bring warmth and laughter into their home. He genuinely does not believe he would be at this point in his life without the foundation and love they provide. This foundation has allowed him to pursue his goals and ultimately created the enduring bond that defines their family.

From his story, we understand the importance of reaching out to those who struggle in the darkness. This was a theme I heard from a number of men who attended the reunion At a critical moment, one person made a critical difference.

Kenn Boles represented Bravo Company as the 1/5 planed for a reunion. Like Ryan, Kenn Boles has a story and like many of the 1/5, his story takes a dark turn. Ultimately, he found redemption and a new mission.

His mother named him Kenneth and called him Ken. Thinking this was too short, he added another 'n' to his nomenclature. Born in the central valley of California, Kenn was the fourth out of sixth kids and labeled the 'smart one.'

He was smart and excelled at mechanical tasks, but he didn't enjoy school. Yet, he drove his siblings to school every day and ran track in the athletic program. As his mom worked long hours, he became responsible for many of the household chores. By the end of his junior year, he had accumulated enough credits to graduate and partied much of his senior year.

He would later learn that he was considered a bad influence by his friends' parents. He excelled in the athletic program. As his junior year drew to a close, the counselor informed him that he was in the top ten percent of his class but he had not taken any honors courses, limiting his options. With sufficient credits to graduate, he enrolled in the VOTECH program his senior year, applying his mechanical apptitude.

He told his mom, "I'm going to enlist in the Marines. I'm 17 years old and can enlist now. Would you like to be involved or am I to do this on my own?" In August 2011, he shipped out to MRCD in San Diego. Boot camp proved to be everything he wanted.

"I knew I had one shot at this. I thought, don't screw this up!"

MCRD is located next to a Marine flight training facility. On September 11, he remembers how quiet the day became without planes landing or taking off. Late morning the drill instructors rolled a television onto the quarter deck. The men watched the two planes striking the Twin Towers in New York and Kenn remembers thinking, "Wow! This is really happening now!"

He served three deployments with Bravo Company in the First Battalion in the 'shit hole', as he describes Iraq. Without a doubt he says he would do it again as he made some awesome friends in the battles they fought together. During one stint in Fallujah, his squad from Bravo Company moved over to fight with the 1st Platoon.

His squad was part of the cordon set up across the southern end of the city with the first objective to take control of a soda factory which would be the staging area for an assault on a mosque that seem to be the center of the insurgent activity. Kenn led a fire team in the assault on the entrance to the soda factory as they employed a rocket launcher to blow away the door. He was exhilarated watching the door implode. Demolition enlarged the entrance and the Marines moved cautiously into the factory.

He experienced gun fights at close range, with combatants at times as close as five feet apart. He killed his first insurgent and remembers thinking, "Now I know I can take someone's life. This was a big threshold to cross.

The 1/5 was a Regimental Combat Team. We had trained together for so long in such depth, very few units could compare. We knew we could crush it. As we moved into Fallujah we were told to stand down. We couldn't believe it. The order to stand down left us really frustrated."

After six months, their deployment in Fallujah ended. First Battalion returned to Camp Pendleton, believing they were done with Iraq. All too soon the 1/5 learned that they would deploy once again, this time to the city of Ramadi in Iraq. Kenn had eight months left in his enlistment but he didn't want to leave the men he had fought with over two previous deployments.

With the third deployment, he became a squad leader with the Second Platoon. Kenn remembers Ramadi as being a nest of improvised explosive devices with combat-experienced insurgents that practiced the art of guerilla warfare. The insurgents they were fighting were combat-experienced from the fight for Fallujah.

The demands on the heavy equipment outstripped the time required to repair the equipment. His technical abilities and skill set were useful as his squad became experts in repairing the damage to the heavy weaponry.

In August 2005, he remembers sitting over a cold beer in a bar one afternoon when he learned that his close friend, Jeffrey Starr, had been killed. In that moment, he felt as if he did not have the right to be happy. The survivor's guilt lasted four years. He drank, he moved from job to job, sleeping poorly - a trauma survivor.

When he learned that one of his company's machine gunner had been shot in the leg which was subsequently amputated, he questioned once again how he had the right to be happy. Alcohol helped dull the pain and quiet the guilt over surviving when others had not. He was drinking more than he was earning. Two months later he was pulled over and received a citation for driving under the influence. Kenn knew this was not a good look for a Marine - the few, the proud, the chosen!

Throughout this time, he remembers people hounding him to get help, to go to the Veterans Clinic. He walked into a Veterans Center and told them, "I need to talk to someone, now!"

At first, his diagnosis was hearing loss which fell far short of explaining why he was in deep trouble. In time, the diagnosis changed to an injured right shoulder, then it was the left shoulder. He was a bit confused about which shoulder had been injured but either way, not surprised at the damage. He had been firing a shoulder-mounted rocket launcher.

Living in Denver, he met two brothers who introduced him to heavy metal rock.

"In the darkness of the music, I came to an understanding of the violence I witnessed in Iraq. I had reached a point where I was as dark as I could become. I didn't think I could get much deeper.

Four years after the death of Jeffrey Starr, Kenn found the courage to visit the Starr family, to grieve with them over the loss of their son.

"I finally started to deal with the trauma," he says. "I was around people who weren't judging me and I could deal with this stuff."

Kenn knew he needed a plan for the rest of his life and considered working as a forest ranger or fire fighter. But he didn't want to spend the next four years in school so he enrolled in a two-year program, studying fire science. He also took a motorcycle maintenance course. He had driven by the Bonneville Salt Flats, stopping to look around and peer across the salt flats from the starting point. In that moment, he decided he was going to build his own Kawasaki motorcycle. After all, what Marine wouldn't love the lure of moving at high speed across the hard packed salt? Years later, after moving to Colorado, building the bike became his mental therapy as he began to work through much of what had happened in his life.

He began working out in the gym and running long miles to earn a slot on a crew fighting wild land fires. The job is rough, as men worked quickly to grub out brush and cut trees to form a break before advancing flames. The smoke and dust, the urgency served both as an outlet for the tension vibrating along his nervous system and a physical challenge for his body.

As his mind and body healed, he realized that he didn't want to live in a large metropolitan center. He began looking at the smaller communities west of Denver. He believed that the mountains and big trees would be a better environment as he healed.

The day came when Kenn got a call for an interview with a Wildland Fire Crew made up of military veterans in Klamath Falls, Oregon. With hard work, the crew qualified as the first Veteran Hot Shot Crew in the United States, responding to "some of the biggest, baddest fires that blazed across the Western states," he says.

In 2022, Kenn left the Wildfire Fire crew and moved back to Colorado. He bought a home and hired on with the Bureau of Land Management to fight fires. Today, he is the captain of an engine crew.

As he considers the road he has traveled through life, he speaks very clearly about what happens when a man is discharged from the Marine Corps. "In that moment, I lost my sense of purpose, my direction and my identity," he says.

This sense of loss underlay the depression he sank into after leaving the Corps. He has worked to bring meaning back to his life in place of the alcohol he used to mute his pain. He has found a new sense of identity. During his off season when the fire danger subsides, Kenn turns a wrench, working on cars and motorcycles. His happiest moments come at the end of the day, sitting in a lawn chair with a cold beer, watching the sun set while dinner sizzles on his grill.

Many of the men leaving the Marine Corps experience the loss Kenn describes. One day, the man knows what is expected. The next day, what he knows and understands is pulled out from under him with no direction ahead. He no longer has a mission, a purpose. These men are poorly prepared to succeed as civilians. Until recently, the Veterans Administration failed to recognize the need for counseling and for intervention in the struggle to assimilate as the Marines move from one culture to the next.

Both Kenn and Ryan each reached a point where they ran out of options. Depression and anger settled in. The memories of the battles, the explosions and trauma became overwhelming. Each man received help unexpectedly to rise above the destruction haunting him. The same can be said for many of the Marines that returned from World War II, from Korea and from Vietnam. Those who do not find that new purpose can be found in bars, on street corners and in quiet graves. Unless another Marine comes along and shares the grief they experience and gives them a hand up, many veterans struggle to redeem their lives after combat.

This reunion was intended to be more than a sharing of stories. The committee hoped men who were struggling would find a link to those who had rebuilt their lives and be encouraged to move forward.

Chapter 6

1st Sergeant's Hill

Years ago, when payday rolled around for the Marines at San Mateo in Camp Pendleton, the sergeant would tell his men to meet front him at 1st Sergeant's Hill to collect their paychecks. He neglected to say that he would be waiting at the summit rather than the base of the mountain. First Sergeant's Hill stands 700 feet in elevation, the trail climbing the steep northern exposure.

Before sunrise, 0600 hours, we were to meet the buses that would transport us to the trail head on Camp Pendleton. One by one, we climbed onto the bus. Among the crowd, I noticed a man who had been severely burned. Scars covered his head and neck, his arms ended in two metal hooks. He took a seat near the back of the bus.

As the bus rolled forward, he yelled, "This is a safety alert. Please keep your hands inside the vehicle at all times!"

Then, thrusting his hooks into the air, he yelled, "Oops!"

The bus exploded in laughter with the Marines shouting, "Oorah!"

After passing through the checkpoint at the gate, Marines and their families arrived at the base of the mountain called 1st Sgt's Hill. A chalky scar, near vertical in places, marked the trail to the summit. Was this our route? One veteran darted through the crowd, moving briskly on two legs, one of flesh and bone, the other a prosthesis.

"Hey, I'll hike with you," I called.

"I'll see you at the top!" he countered.

Quiet conversations around us mentioned the use of chains to assist in the climb. We were directed away from the training route to a less demanding dirt track. The men moved out dropping into a wash at the base of the mountain, only to be confronted by a stream bed with water rising up to their knees. A dry wash is not much of an obstacle. This year the rains had been heavy. A week before, the water had been waist high. No bridge, no stepping stones. Two Marines pulled their wives onto their backs and forged ahead. As each hiker considered the choice to move forward or return to the parking lot, the group began to splash across the stream.

A handful of Gold Star parents remained in the parking area, gazing upward at the summit, wistfully remembering the days when we would have climbed along with the Marines.

"C'mon, over here!" Sergeant Major Luke Converse led us to white vans. We rolled south along the base of the peak before turning toward a bridge over the flooded river bed. The track rose steadily as the passengers joked about whether the van was capable of climbing the steep grade. Up, up, up! Topping a saddle, we plunged downward, only to steeply climb again. At the top of the hill, I caught sight of the first crosses. As we walked forward, our descent revealed a forest of crosses, each a quiet testament to the service of our men in uniform. The sense that we were standing on sacred ground crept over us as we silently read the inscriptions.

No hot house bouquets lay at the foot of these beams. No, a meadow of wild flowers paid homage to the memorial, disguising the rough volcanic rock that cluttered the hilltop. Some crosses bore shredded flags, a pair of weathered boots lay at the foot of another cross. One stood in a mound of empty beer cans, speaking of a wake in honor of those who had given their lives. The crosses seemed to stretch on and on, along the path.

In the midst stood one cross stood hidden by camouflage netting, waiting the moment it would be revealed.

Those hiking to the summit began to appear, two or three at a time until the hillside around the shrouded cross was packed with Marines. Among them, the Gold Star families and those of the Fallen stood gazing upward, awaiting the moment when the shroud would be torn aside.

I glanced around to find Marines quietly talking and hugging family members as they recalled incidents in Iraq, releasing memories. Some bowed their heads in silence, a few wiped away tears. The official memorial stands at the foot of 1st Sergeant's Hill. This hilltop memorial has been erected by the grunts, the Marines who serve in the infantry ranks. As they returned from deployment, they carried the crosses to the summit, consecrating them to those who are no longer present.

As we gathered near the shroud, I realized I was standing next to Major General Fred Padilla. General Padilla had been the commanding officer of the 1/5 as they crossed the Iraqi border. My son had spoken of him with the utmost respect, telling us that this commander cared for his men. Earlier, I watched as he moved through the crowd, joining the conversations that flowed around us as if he were one more Marine rather than the one who knew the responsibility of command. Even with the camaraderie I could see the respect of the men for their former commanded. I hesitated before speaking to him. I knew little of his character but one question had haunted me from the day my son had first announced his intention to become a Marine. As a mother, I wondered if the officers would value the life of my son?

Seizing the moment, I turned to the General to introduce myself as the mother of Lance Corporal Marty Mortenson, one of the men who had been lost in Ramadi, Iraq.

"Sir, when my son enlisted, I had one concern. I've read about the battles of the Civil War and World War I, about the men who died. In one charge their bodies pushed down into the mud by the feet of those who scrambled over them. I've read about officers who ordered assaults that were sure to fail. And, as I thought about my son in the Marine Corps, I wondered if his officers would value his life as much as I did?"

And then I said nothing, awaiting his response.

After a moment, Major General Padilla began to talk about the moment when the 1st Battalion Fifth Marines crossed the border between Kuwait and Iraq. They sat in the desert for weeks while politicians bickered in an effort to forestall a bloody conflict. Words came to nothing. The 1/5 would be the first to cross the border with orders to take the southern oil fields before the oil wells could be set ablaze.

The evening before the campaign began, Major General Fred Padilla rehearsed the plans with his commanders. They were scheduled to cross the border at sunrise but as surveillance detected the movement of Iraqi forces moving into the oil fields, the Central Command feared that they would attempt to destroy the oil wells. The 1/5 was ordered to move forward immediately.

Engines revved to life, men shouldered their packs. Squads of Marines entered the amphibious assault vehicles that would each carry as many as 25 men northward toward Baghdad. Their nerves tightened. This was the moment they had anticipated, no further delays.

The mile-long convoy roared to life as vehicles crept forward through razor wire. Grunts may be the fighting force of 1/5 but the convoy was accompanied by three Abrams tanks, and teams of support personnel, combat corpsman, chaplains, and those who handle logistics for a battalion on the move. Many were riding in light armored vehicles and canvas-sided trucks, all en route to capture the oilfields.

As they moved forward, they encountered hundreds of Iraqi soldiers raising their arms in surrender. All too soon the first artillery round roared toward the convoy and the men of Alpha (Apache) Company were returning fire. One bullet struck Lieutenant Shane Childers. As he bled, the corpsman fought to keep him conscious until the Medevac team arrived. It was a losing battle. General Padilla received the call and arrived shortly after to the tear-stained faces of the corpsman that had sought to save Childers.

As we stood at the foot of the cross, he recounted his grief and apprehension at the loss of his officer. That evening, he wondered how many men he would lose in combat as they moved toward Baghdad. And then he stop speaking. We stood there for a moment in silence.

"Thank you, General, for listening to me."

"Thank you for talking to me," he replied. We said nothing more, waiting for the ceremony to begin.

My respect for the man no longer hinged on my son's words but in this man's demeanor. In a sense, I had challenged him. He could have been offended but he recognized that I longed for an answer. My question was one that men have pondered through time immemorial. What is the life of a man worth?

Our attention was pulled back to the cross as the short dedication began. Men pulled the netting back to reveal a cross even as a sea of arms stretched upward, many of the men holding their cell phones up to record the moment. The speaker reminded us of the cost that had been paid in defending our values and the fight for freedom on other shores.

Chaplain Carey Cash, the chaplain who had traveled with the 1/5 as they advanced into Baghdad, was called on to dedicate the memorial in prayer. For these men, this cross is their cross, unique from every other standard on that hill. They will gather here in the future. Indeed, the following year, they gathered again at the foot of the cross to remember those who had gone before.

As the dedication ended, a cry of "oorah!" rose from a hundred throats. Gold Star families were summoned to the cross to claim the replicas of their son's dog tags. Quiet conversations drifted across the hilltop as men traded stories of their service. My husband and I stood looking out over the countryside, so quiet and yet so vital to these men who had once trained to carry the fight into Iraq.

One of the Marines approached us and we commented on the view. He revealed that he was the current commander of the base and had joined the group climbing the hillside. He pointed out landmarks and the routes the men ran as part of their physical training. We laughed about the river at the base of the hill, ruing the best laid plans and the shock of cold water entering their boots.

Slowly, the group dispersed and returned along the route they had climbed. The Gold Star parents returned to the vans, descending to the Parade Deck on the base below.

A second ceremony was held on the parade deck. General Fred Padilla addressed the men he had commanded and they listened carefully as he spoke of their bravery and persistence. This reunion was not just a recitation of the battles fought as the battalion moved toward Baghdad. This was a time to draw together, to remember that we are not alone in the battles we encounter in life. We fight together, we support each other. The Marines are taught in training that they fight for the man on their right, the man on their left. Through this time of drawing together, the men hoped to re-establish the comradeship they once shared as they faced enemy combatants together.

When the service was dismissed, the Marines and their families moved across the parade deck to examine the weapons that these men once carried in and out of battle. Other weapons from previous eras were also on display. Fathers showed their sons and daughters the weapons and how they operated. Some lined up the assault rifles on distant targets.

I glanced over to a mother who had place her 15 month-old child behind one weapon, his fingers playing with the knobs and trigger. She laughed, looking in our direction as if to say, 'Look at my future soldier.' In the pride and pain I felt over my own son, in knowing the pain of loss, I turned away, unable to share her amusement.

The monument to the men of the Fifth Regiment lost in combat stands a short distance from the parade deck in a grove of eucalyptus. Passing through wooden arches, visitors arrive at a wall lining the back of a plaza. The surface is engraved with the names of men in the Fifth Regiment lost in our country's service. A second stone stands in the center of the memorial with names more recent in our history. Around these are the memorial stones of other conflicts as well as a stone from the country of Vietnam, thanking the Marines for their service. All are somber reminders of the cost paid for freedom, whether on our shores or in lands foreign to our warriors. The silence is broken by the voices of those who scan the walls for the names of those they knew well. Some are family members and there are veterans recalling the men with whom they served.

Lifting one's eyes beyond the monument, the trail up 1st Sgt's Hill stands witness to the nascent of these men's service while these stone monuments speak of sacrifice. The narrative caught between nascent and abrupt closure is written on the memories of those who visit. First Sgt's Hill is a tangle of pride, of training and of service relinquished.

My thoughts returned to Major General Fred Padilla, a man who shouldered the responsibility for an entire battalion of men.

The responsibility lies heavy on a commander's shoulders. The men do not easily grant respect to an officer. The officers understood that theirs was the responsibility to lead and to train on this site in preparation for the opposition they would encounter.

Over the past four hours, I had watched General Padilla. He easily mixed with the former infantry. To my knowledge, none remained in active service and the General encouraged them to address him as Fred, rather than Sir. How did he come to the moment, I thought. Who is this man that once commanded 1,200 Marines?

Chapter 7

A Measure of Respect

When a man considers service in the Marine Corps, he may be drawn to infantry or to command. The two are separated by more than a title as most officers do not enlist or train alongside the grunts. There are three routes to becoming an officer. A young man might receive a recommendation to the Naval Academy from a congressman. Or he might enroll in a Reserve Officers Training Corps (ROTC) program while seeking a four-year degree at an accredited university. A third route is open to an enlisted man who is chosen to enter the Officer Candidate Program. These three routes funnel men into Officer Infantry School with the competition tight for the open billets.

When I first entered the reception in 2022, I met Lt. Colonel Kelly Thompson, Retired. One of the grunts nodded at his former commander, saying, "Yea, we like him! He was a good officer."

Not every officer receives such an endorsement from an enlisted man. In talking with the non-commissioned officers, I learned that Marine Corps officers are vetted in their leadership skills by both their peers and the enlisted men. Course work is encouraged to allow the officers to develop their skills.

The Marines of the 1/5 respected Thompson for his commitment to lead his men from the front. He was unlikely to hang back at headquarters while they went out on patrol.

Josh Shores recalls an incident when a close friend had been killed. That evening, while standing watch, Thompson, his commander dropped by the post to sit awhile and listen to the young Marine's grief and confusion.

Not every officer shared that commitment. In a second incident, another enlisted man related an incident involving a junior officer the men did not respect. As the squad climbed into the transport to begin their patrol, this officer remained back at headquarters. His squad dropped into a neighborhood in a location near a vehicle with the back end sagging toward the ground. This was an indication that the trunk was loaded with explosives and wired to ignite. The men repeatedly called their commander, requesting the explosive ordinance team destroy the vehicle rather than leaving them exposed to the risk of an explosion. Despite repeated requests, the officer their judgement and delayed acting, leaving his men feeling exposed.

I looked forward to interviewing General Padilla to learn what he valued in leadership but health concerns delayed that opportunity. He was previously interviewed by Jeremy Stalnecker, the CEO of Mighty Oaks Foundation regarding the principles of leadership, giving us a view of the General from one of his officers during OIF 1.

Fred Padilla was the middle child in a family with five children, including one younger brother and one older brother. He describes himself as one who didn't know how to quit, picking on his younger brother who then called the older sibling to help him. The older brother jumped into the fray as the younger brother ran for their father who quickly settled any dispute. Fred caught his father's discipline as he crossed the line, failing to understand that to quit can be either detrimental or an asset.

Fred describes himself a an average kid who was observant, watching what occurred around him. His father was a fighter pilot with the Air Force and the family lived on Air Force bases where Fred observed the treatment received by the men returning from Vietnam. He saw the disrespect they were accorded and felt the show of contempt was wrong for men who had been willing to sacrifice their lives. This inspired him to serve his country and he began to consider a military career.

His dad asked what branch he was considering? Both his brothers and his two sisters would serve in the Air Force and Padilla thought he might go that route. One problem, he had tried flying and he didn't like it nor was he good at it. His dad suggested he consider one of the other branches of military service.

"I thought about the Army. I became a Marine because the Army recruiter was so bad. After visiting the Army recruiter, I went to the Marine Recruiter and they were sharp. I was so impressed. I didn't know much about the Marines but I knew they were smaller than other branches and the uniforms were really cool."

After training in Infantry School, Padilla's first billet was with the 3rd Marines, 6th Battalion, followed by the 9th Marines as the commander for Weapons Company. In those early years, he recalls serving alongside some commanders who were outstanding examples of leadership. One of those men was Steve Carpenter, an electrical engineer out of the Naval Academy.

"He was an enlisted guy who became an officer." recalled Padilla. "He was an interesting guy and had gone through Ranger School. Steve was a country boy who loved to hunt and was 'awesome in the woods.'" Carpenter mentored Padilla, teaching him how to plan and implement an event.

"He would sit down with me and make sure that everything was tied to learning objectives. Afterward, we would evaluate the training and remedy where needed. His teaching was invaluable."

"It would have been easier for Carpenter to do this all himself but instead he sat down with me and taught me how to do this. Mentorship versus heavy-handed leadership is much more inspirational and brings out the best in others. This inspired me to teach others."

As he reflected back on the mentoring he received, he added, "Your job as a leader, the final report card, has yet to be tallied. You won't find out the final impact of your leadership till years down the road. How did those Marines you led do in the long run?

"I came into the Marine Corps with an attitude of being uncertain of how this would all work out. When I got to the fleet, I remember thinking this is where I belong. I'm home!"

The training would become invaluable when in 2001, he was sent to the 1st Battalion, 5th Marines as Battalion Commander. A month later, two planes hit the World Trade Center towers.

"We had a lot of advantages.' he recalls. "We had time to train. The second advantage was that we had some freaking, awesome people, just some of the best Marines from the top to the bottom. All of them just awesome. We needed time to develop a cohesive team."

His men would later credit him for pushing them to train until every move was ingrained in their thoughts and actions. That training would become invaluable on April 10, 2003 as the 1/5 entered Bagdad and were engulfed by a fire storm. Padilla witnessed how the hours of training paid off.

"When you have something like April 10 happen when just about everything that could go wrong does and every Marine knew what he had to do. It wasn't because of me as the commander, everyone did what they had knew they had to do. To be in a city a big as Los Angeles at night, no lights with night vision equipment, under fire the whole time and we made our three objectives. Ultimately we secured the objective that was assigned to us. We had vehicles all over the place and it all came together at the end.

"How did that happen? It was a bunch of Marines that knew what had to happen and did it despite the significant hardship and the overwhelming uncertainty. I am certain there was no shortage of Marines that were feeling their own mortality at that time. This was one of the most satisfying things I have ever experienced in my life."

The 1/5 lost Gunny Jeffrey Bohr in the firefight and others were badly injured. Bohr was the second Marine to die from 1/5. In talking about the men killed, Padilla reflected on the loss of Lieutenant Shane Childers twenty days before they entered Bagdad. The 1/5 had crossed the border into Iraq. He remembers being frustrated as a line of tanks kept moving off the designated route and each time he corrected their course over the radio. Finally, he moved to the front of the line and directed them to follow his humvee until they reached a well marked lane through the mine field. As they cleared the mine field, they entered their first engagement, coming under fire. A report came over the radio of a man down.

"When the report came over the radio, I had a sinking feeling in my stomach. We were on our way to that location and I told my driver to step on it. Then I heard the call that it was routine and I knew that Shane was gone. When I arrived, I saw the faces of the men. Tears were streaming down the face of the medic, blood up to his elbows. He told me they tried to save him but couldn't. Shane had been shot in the liver, he could not survive that. I told the corpsman he had done his best.

"I knew I had to talk to Shane's Marines and I started down the line. They were all asking, 'How's our Lieutenant? I had to tell them he was gone.' At that moment, I knew I was the old guy and asked myself, 'What is my job?' I knew I had to help my Marines refocus. I knew they were all hurting real bad. I had to help them realize that bad things happen but we have to finish the mission. I told them, you have to go on and finish as he trained you to do. That night, I went into my space beneath the camo netting. I allowed myself to feel the loss. I sat there, wondering, 'Will there be more of this?'" Padilla talks of reaching into his mental toolbox when challenging times like the loss of his men confront him.

"You've got a lot of tools in your tool box to meet the challenges. Some challenges will hit you in the mouth and knock you on your ass. It is going to be hard to get up. You reach into that toolbox and grab those tools.

"The number one tool is faith. I don't know how anyone can be resilient and leverage that kind of strength without faith. I have relied on faith (in God) my entire life." he explains. "Faith keeps you grounded. It helps you from spinning out of control. Faith gives you clarity of thought."

In his faith, Padilla find hope, even in the moment when an event turns ugly.

"Hope keeps you going when you lose a man in combat. We should not be afraid of living our faith."

He reminded his men that courage rises out of being steady, of not giving up and in choosing to do the right things. Being steady rises out of faith and is part of his foundation as he considers his service to family and country. He uses an acronym, F2C2, which stands for Faith - Family - Country - Corps.

With a smile, he says, "It doesn't always come in that order but these four are important. I serve the Corps through my faith, the same with my family. That doesn't mean it works out equally. I don't get quantity of time with my family while I'm serving in the Corps. I make sure my time with my family is quality!"

A second tool in his box is humility and this goes hand in hand with the ability to mentor others.

"This is not about you or me," Padilla says. "We are to be servant leaders. Humility is such an important part of leadership."

He explains that a Marine lives by Corps values: Honor, Courage and Commitment. I've heard this from other men who served in the 1/5 under General Padilla.

Training, faith, courage, and hope formed the basis of Padilla's leadership and the measure by which his men believed in and respected his leadership. When the 1/5 returned from their first deployment into Iraq, General Padilla was assigned to a new billet. He would later say he missed talking with the men he had commanded in 2002. The reunion was his opportunity to reconnect with the Marines of the 1/5 and hear how they were transformed by their experience.

As I think about the principles of leadership, my thoughts returned to Lt. Colonel Kelly Thompson that first evening. He stood with two other men, holding a drink in one hand and quietly watched his Marines catching up. When I learned that he had been the Alpha Company Commander, I asked if he remembered my son. He nodded with a quiet smile, saying, "Marty was great."

Nothing more. I stood there, mute, uncertain of what to ask next.

"So, what can you tell me about him. I know my son. Really, how was he?"

Again, the quiet smile and the reassurance, "Really, he was great." I sensed I would learn nothing more. Every mother yearns to hear more about her boy.

Months later, as I began writing about the these men, Sergeant Major Converse casually commented that Thompson was "a great officer and a man who suffered greatly as he grieved over the men he had lost and the trauma of combat."

Kelly Thompson was no stranger to grit and heat after growing up in west Texas. Fallujah is a long way from Thompson's home town of Shallowater, Texas. Each year, intense sun and weather assault the roofs, of his home town, providing regular employment for the young men of this town of 3,000 people. At age 17, Thompson straddled the ridge line of a roof, nailing down shingles. Looking around, he could see the future on windblown rooftops as men in their 20's worked alongside him. Kelly wanted more out of life than shingling roofs.

He marched with his graduating class of 50 young men and women before catching a Greyhound bus south to Lubbock. When he first enlisted, he was told the Marine Corps didn't need him. The Corps was in a stop-loss phase as they allowed the number of enlisted men to drop by attrition. He was offered the option of enlisting on an open contract which meant four years of active duty, followed by four years in the Marine Corps Reserve.

His first deployment was as a Corporal in Desert Storm. Four years after he first enlisted, he moved from active duty to Active Ready Reserve.

"After Desert Storm, boom, we got out unceremoniously. We all were like broken toys. I knew I wanted to go to college, not to just earn a degree but to be educated. I chose the University of Texas in Denton which was a five-hour drive from my home town. I was the second one in my family to graduate from high school, the first to earn a college degree."

Thompson chose to seek a Commissioned rank through the ROTC program offered at the University of Texas. During the summer, he attended the Platoon Leaders Class in the Officer Candidate School, knowing the course had a 50% washout rate.

"August of 1995 was memorable. I was commissioned as

a 2nd Lieutenant in the Corps and my wife divorced me," he recalls. "I moved on to six months in the Basic School which trains and develops new officer candidates. It is very competitive with 250 candidates for ten slots."

Thompson graduated from Basic and moved on to the Infantry Officer Course which has a high wash out rate. Upon completion, he felt he was in good shape as a newly commissioned 2nd Lieutenant. There was no guarantee for a particular position upon finishing the course and he competed for a slot with the Infantry. He was successful and moved to New Officer Course, becoming a Lieutenant with an assignment to Alpha Company as a Lieutenant in 1996. Three years later, he returned to Coronado Island to teach reconnaissance and boat company skills. He had qualified as a combat diver and parachutist. I was beginning to understand that the training never stops for the Marine officer.

As the 1/5 moved into Iraq in 2003, Kelly was enrolled in the Staff Captain's course with the Expeditionary Warfare school. He gained more training in aviation, supply and logistics, engineering and intelligence.

The 1/5 returned to Iraq in 2004 and Kelly Thompson was promoted to Captain, assuming command of Alpha Company after Lt. Colonel Philip Treglia returned stateside. Lt.. Colonel Bryne had been tapped to lead 1/5 on this deployment. Setting up the cordon south of Fallujah, Alpha Company seemed to encounter the toughest fights and men were getting gravely injured or killed. Thompson's approach with Company A began with leading his men from the front in each patrol even as he thought he was going to die. He says he used "to pray for strength and wisdom and the ability to be a good officer."

He describes Fallujah as a mess, they were killing bad people. No one wanted to go out on patrol with Alpha Company as we encountered some of the toughest fights. Men were getting gravely injured or dying."

Every time Alpha Company paced through the streets of Fallujah, every time the men rolled out in convoy, Thompson led from the front. His was the first vehicle out the gate of their camp. He never knew when an RPG would fly or bullets would pound the sheets of metal welded to their vehicles.

He later described his irritation over OIF 2. We were in full combat. When the national media got hold of the story, ordered to stop. This was so stupid. We could have taken the city if we had been allowed to move forward. This gave the insurgents time to fortify and build up their positions. After we left, the 3/1 had to resume the fight all over again."

Throughout his first two deployments, Thompson became convinced he was going to die. He could not escape his destiny. Then, he found himself on a plane, returning home, in one piece! He had survived! He began to consider his next assignment. Nearly six months later he learned he would be returning to Iraq once again with the 1/5.

"The profanity flew." recalls Thompson. "My vehicle had been destroyed five times in Fallujah but General Eric Smith had argued with the Division commanders to retain me as Commander of Alpha Company."

As Alpha Company entered Ramadi, they were assigned to the Government Center, taking up the fight in the middle of a functional city. Unlike Fallujah, which had been non-functional with the residents fleeing, Ramadi was a miasma of tribal governments and Shia rule. The city lies west of Fallujah with the capital of Bagdad further westward in the triangle of the Baathist home ground. The three cities, along with al-Najaf, formed a hot bed of insurgency after the invasion of Iraq. Negotiators might meet with one sheik only to be undermined by another. The insurgency was now led by professional fighters, including battled scarred men from Chechnya, a former Republic of the Soviet Union.

"In one way I was glad I returned. I am convinced to this day that there would have been more Alpha Marines killed had I not been company commander because first I was judicious in my decision making and second, because I went first. When you're going first, you're not going to make stupid decisions.

"I used to pray. I didn't pray to live. I prayed for strength, for wisdom, and the ability to be a good leader and not a coward. I knew I was going to die."

The fire fights were often at point blank range, as they came out of ambush at us or we came out at them. You had guys like Lance Corporal Browne killing the enemy with his nine millimeter from a gun turret because he couldn't depress his machine gun low enough to fire on the men coming at him. This was a war of ambush and hidden improvised explosives. We knew we had to kill or we would be killed."

Thompson led his men out on patrols, knowing at any minute the ground could erupt from buried explosives or an attack erupt from under the hood of a car or the windows of a silent building. "

When an IED explodes on your vehicle, it knocks you out. The blast pulls the air out of your lungs and you look like you have died. You wake up and you're alive. You think you're okay."

"The most difficult decisions were how to accomplish a mission and still keep the men alive. My thinking was a wholistic approach, examining my orders and finding the best way to accomplish the task without harm to the men."

Thompson entered Ramadi with 174 men. He lost nine and 33 more were wounded in their six-month deployment.

"As the six month deployment came to an end, Thompson returned to Kuwait to catch a flight home. Iraq was behind him and he remained alive. He stood on the Parade Deck before Alpha Company at Camp Pendleton, watching the families reunite with their men, even as he knew other families looked at him, holding him responsible for the deaths of their sons. He had done all he could to protect the men of Alpha Company so ultimately, he remained silent.

Six months passed. He was serving with NATO in the Netherlands when he became convinced he was suffering a heart attack. Medical help did not relieve his symptoms. He was diagnosed with post traumatic syndrome. As a single man living alone, he did not have anyone to talk with as the anxiety and nightmares brought on depression. For two years, he lived for on edge, white knuckled, hoping the PTSD would pass.

"It was a very lonely time," he quietly acknowledged.

Another eighteen months passed and Thompson received a call from General Eric Smith, the 1/5 Commander who had retained him for a third deployment to Iraq.

"What's your next assignment?" Smith asked. "Where do you want to go next?"

"Sir, I request an assignment near a hospital where I can seek medical treatment," Thompson told his former commander.

"I had experienced too many explosions. I was diagnosed with a traumatic brain injury and PTSD. If you get blown up multiple times, your brain is going to get injured! My memory was not the best. I needed help. With a hospital next door, I could seek treatment."

Through counseling, he talked and talked some more about all he had done and the results of his decisions.

"I talked about it so much, I became sick of talking about it." I took medication to assist with the anxiety."

Ultimately, he understood that he would just live with the memories or as one veteran describes the past, he lives with the ghosts of war. Leaving active duty, he understood that he must find employment to keep his mind from replaying the trauma he experienced. Yet, there are times when he turns to alcohol to numb the pain.

Not all officers lead from the front but many of the men who take pride in "the few, the proud, the Marines" will tell you that the responsibility of command requires they lead rather than follow. The price is high, the responsibility unremitting. Thompson believes he still leads his grunts in setting an example by seeking help when he came to understand that he could not pay the price of experience alone.

Now retired from active service, Major General Padilla and Lt. Colonel Thompson are men who stepped forward believing that they had the ability to lead other men in serving their country. They chose to accept a commission to lead. The two elements of leadership that stood out in my conversations with them were the willingness to lead from the front and in mentoring those they command rather than simply issuing orders. I admired their ability to listen to their men.

There is another roll call of officers who rose out of the enlisted ranks. Men equally committed to those they command. These non-commissioned officers rise among their fellow grunts, stepping into the chain of command between those designated for leadership from the moment they sign and those who carry the mission on their backs.

Chapter 8

To Command

Where does the training for command of a Platoon or a Battalion begin? Is command grounded in the training of an officer or among the 'grunts, either by example or by designation? When a man enlists and is admitted to the United States Marine Corps, the smallest of the Armed Forces, he enters as a recruit. The first 13 weeks are difficult with Hell Week, known as the Crucible, draining the recruits of every internal resource a man can summon.

When the recruit arrives at the intake center, he steps into the chair where his head is shaved. As he walks away, his hand rubbing the nubs across his scalp, he sees other recruits with the same sheepish expression. They realize they all look the same. He removes his clothing and is assigned his first set of camouflage, which he will wear for the next 91 days. If he thought sleep was an ally, he soon learns that the lights go out early and the call to rise comes even earlier. Sleep deprivation steals over his senses and he find the physical challenges ever more difficult as his muscles scream for adequate rest.

What is the purpose of this form of institutionalized abuse? The recruit is stripped of everything he has adopted to define who he is as a person. His ego is worthless. He responds to each and every inquiry as 'this recruit, sir.' The word 'I' is not permitted when speaking to a drill instructor.

After 13 weeks the recruits have been stripped of their individual identities to become hardened, disciplined members of a team that can function as a unit when the bullets and rocket-propelled grenades are flying. Without thinking of personal ambition, the men move forward, instinctively fulfilling the mission they have been given. Even after they have been discharged from the Corps, most Marines do not lose this sense of discipline.

Passing through the ranks of Private, Private First Class and Lance Corporal, a Marine may rise to the noncommissioned rank of Corporal, commanding a fire team, followed by Sergeant who leads three fire teams. Each promotion is based on a man's performance. This is the underlying strength of the Marine Corps. A Marine is promoted based on his performance and is responsible for training his subordinates. He is held accountable for their actions as a squad. He is to enforce military standards and assist in the professional development of each man in his squad.

With the next promotion, a Staff Sergeant commands a Platoon and becomes the liaison between the enlisted men and a 1st or 2nd Lieutenant who directs a platoon. When a Lieutenant, a commissioned officer, take command of a Platoon, the staff sergeant begins by introducing each man along with their background, their families and their record of service. How does a Staff Sergeant knows these personal details? A platoon consists of 30 to 40 men and the staff sergeant learns these details by talking to each man, keeping notes until he has committed each Marine's personal history to memory. He is an integral part of the development of each man personally and in his professional career.

With each succeeding rank, a noncommissioned officer (NCO) is responsible for the men in rank below him. NCOs tend to the training and appearance of their Platoons. They are responsible for the actions of their men and participate in the planning of each mission at their graded level.

Upon promotion a Staff Sergeant becomes a Gunnery Sergeant, known as the Gunny. Moving up from Gunnery Sergeant, a Marine must choose between working toward a promotion to 1st Sergeant or Master Sergeant. The selection process is very competitive. A First Sergeant has command responsibilities and may work to earn the rank of Sergeant Major, the highest rank for an enlisted man. He must show outstanding leadership qualities and professional competence in working independently under the Company Commander. Sgt Major Converse emphasizes the point of a 1st Sergeant having a smaller influence on a large group of men even as a Staff Sergeant has a much stronger influence on a smaller group of men.

A Gunny will seek to be promoted to Master Sergeant, becoming a technical expert in his field, assisting the Company Commander in technical and tactical requirements. He is eligible for promotion to Master Gunnery Sergeant.

When a Marine is promoted to First Sergeant, he oversees 100 to 200 plus men. Upon reaching the rank of Sergeant Major, he becomes responsible for 800 to 1,000 plus men. In that task he is assisted by the noncommissioned officers in ranks beneath him. The ability to command is embedded in these ranks, beginning with the most basic training for the recruit. The NCO learns that discipline is the key to succeed and within this discipline he is willing to follow command. A well-managed Battalion reflects how the roles of each man dovetail and complement each other.

Both Daniel Santiago and Tracy Offutt earned the rank of First Sergeant. Both men know the struggle of combat and leading their men. On April 13, a Special Operations helicopter landed hard in a controlled crash southeast of the 1/5 positions on the outskirts of Fallujah. Lt. Colonel Patrick Byrne responded with a Mobile Assault Platoon (QRF) specifically trained to respond to emergencies. The incident is described by Bing West in No True Glory.*

Driving across farm fields, the team reached the site after the crew had been evacuated. They quickly gathered sensitive gear, a knapsack and a transmitter before settling around the site for the evening hours.

In the morning, the team woke to mortar shells dropping around them. The fields were soft and they quickly realized that they must drive along the tops of irrigation ditches, exposed to incoming fire. Under fire, they evacuated in the wrong direction and their 1st Lt. Josh Glover turned men and vehicles back toward their camp. Along the route insurgents had gathered, targeting the American vehicles with small arms and rocket-propelled grenades. The Marines returned fire with Sgt. Cparski burning through seven hundred rounds of .50 caliber ammunition in the fire fight. Behind

* No True Glory, Bing West, 2005, published by Bantam Book, pg.148-149.

Cparski, a second gunner burned through over a thousand rounds. The squad of Marines, seated on center benches, fired their M-16's toward any incoming fire.

Private First Class Boye was hit in the leg by an RPG. When the blood could not be stanched, 1st Sgt. Santiago screamed over the radio at the drivers to push their vehicles faster. Every vehicle was driving on rims, their tires shredded under the withering fire. They did not reach the aid station in time and Boye died en route from the blood loss.

A First Sergeant may bellow when he sees something amiss or under assault. Santiago will tell you that at other times, he adopts a more persuasive tone, finding a topic that will appeal to his men as he seeks to instruct his Marines.

Daniel Santiago was born in Puerto Rico. As an American Territory, its citizens are accorded the rights of those born in one of the 50 states. Santiago's mother worked for the Federal government and when she was offered a better position in Chicago, she did not hesitate to move her family north to Illinois. The family settled in Chicago. Santiago was nine years of age and did not speak a word of English. When officials at a private school indicated he would be set back a grade upon enrollment, his mother insisted that her sons were as smart as any other child in the school. They would be enrolled in a standard class, learning English through total emersion. For Puerto Ricans, education with an emphasis on science and math is very important. Within one month, both Santiago and his brother were fluent in conversational English.

There were other lessons for the two boys that had nothing to do with school. When they lived in Puerto Rico, their father constructed two shoe-shine boxes, giving one to each of his sons with the instructions that they were to shine the shoes of taxi drivers. They walked to the taxi stand on the corner, certain that the taxi drivers were laughing at them. His father wished to teach his sons humility. He worked long hours as a mechanic at a small neighborhood gas station he owned with his partner. On weekends the family would pass out Christian tracts door-to-door throughout their neighborhood. Christianity was the basis of their family's culture.

"My parents showed me how to live," he recalls. "It made me

a different person. I understand respect. After moving to Chicago, we were sent back to Puerto Rico every summer to spend time with our family. My parents placed their emphasis on education, on sports participation and being involved in your community. I learned that I was a part of something bigger than myself."

He believes that this early foundation made him a better man.

"Today, I understand why the Marines have to break down the recruits and build them back up. Hair brings individuality. We have to recognize that we are all the same. In the Marines, we are created to be a team. Through this process we establish discipline and coordination, just as my family did in my early lessons. I understand that it is an honor and a privilege to serve my country as a Marine."

He enlisted out of high school. His brother would follow him as an Army Ranger nearly 18 years later. With his background, Santiago was destined to seek more than four years as a grunt. After serving four years with Third Battalion Fifth Marines, he was assigned to 1/5. In 2000, he was given his first special duty assignment as a drill instructor. He returned to the 1/5 in 2003, ranked as a Staff Sgt, deploying to Fallu'ah and a year later to Ramadi. Following the deployment to Ramadi, he was assigned to home ground in Chicago as a recruiter. Through both his assignments, first as a drill instructor and later as a recruiter, he drew on his combat experience in talking to his men. We talked about how these two special duty assignments differed.

"As a recruiter you learn persuasion. You learn to talk to these young men about their opportunities in the Corps." He paused for a moment. "Ma'am, I must correct you on one thing. We are not drill sergeants. You write those words, other Marines are going to be looking at me, wondering."

We both laughed and I promised to use the correct term. Drill Instructor! Marines are trained by one of their own, a man who has come up through boot camp and knows the grind of being a grunt.

"As a drill instructor, I had to use discipline. I had to be strict in training the recruits. I learned both persuasion as a recruiter and discipline as a drill instructor. Both of those are important and I used both as a First Sergeant."

Later, thinking back to Santiago's statements about the Marines vetting their officers, about the humility he learned as a boy and his belief that it is an honor and privilege to serve as a Marine. I understood that this was an example of a noncommissioned officer who sought to provide an excellent model to the men under his command rather than one of heavy-handed discipline. The men understand that their drill instructor has come through the same training and knows their struggle to perform.

First Sergeant Tracy Offutt did not begin his military service in the Marines. Initially he thought to enlist in the Coast Guard but when the recruiter failed to follow up, he turned to the Army. Offutt was an Army brat, his family following his dad around European bases before settling in New Mexico. Offutt wanted out of New Mexico! He completed his service in the Army in three years. Eight months after his discharge, he returned to enlist with the US Marine Corps and served 13 deployments, including combat prior to Operation Enduring Freedom.

He had enlisted in a peace-time Marine Corps. After September 11, 2001, everything changed. He deployed on three tours in three years with the 1/5, starting with Operation Iraqi Freedom (OIF). Pursuing one promotion after another, he rose to 1st Sergeant during OIF, eventually being promoted to the rank of Sergeant Major.

Initially, he may not have aspired to the responsibility of command but Offutt knew his Company needed a 1st Sergeant. He had not turned in his paperwork with a photo for consideration as the promotion board was about to meet. His Platoon Commander reached out and told him to get his photo submitted that day. As he rolled south out of San Mateo, he felt as if God reached out, saying, "This is something you are going to do!" He returned with the photo and days later was promoted to the rank of 1st Sgt. Up to this point, he had been with the 3/5 but after the 1/5 Battalion returned from Fallujah, he was transferred in to Alpha Company. It is difficult to step into the position of another officer in a combat-riven company.

"The Commander at every level must know that every word he utters has a purpose." he says. "Every action is an example to his officer and the men he commands. The officers in turn, look to the Battalion Commander as their example. Some may be a little more lenient, others know the struggle to perform."

He explains further than a 1st Sergeant cannot buddy up with the men he oversees. He is aware that they are watching his actions and he is an example of his men. He relies on his Staff Sergeant to assist in planning as the Staff Sgt knows the mental and physical condition of the men. He found his most difficult decisions were in deciding a man's future in the Marines.

"As a 1st Sgt., sometimes you have to decide a man's entire future in the Marine Corps. I couldn't see into the future. I had to determine what a man's potential was. It could be easy to follow the rules but I had to be willing to look beyond the rules at what the man could become. There is an exception to every rule and sometimes a little grace goes a long way. My first responsibility is to the Marine Corps as the institution, my second responsibility is to the individual. There were times when I gave a man a second, even a third chance when he screwed up and I ended up looking like a fool. I was willing to believe in a man to see him develop into a Marine."

Through the Christian faith, he came to understand the basis for giving a man a second chance. This understanding of the grace he had received became an example of showing grace toward another person.

"I was a faithful person but I had no depth, no understanding. I had my practice of faith all backwards, believing that I had to be a good guy to be accepted by God. I learned that it is by His grace that we come to God, not through our own merit."

"I watched one of my men who was just a savage, not a good person. He had no skills for living in a modern society. No one wanted to be around him. He became a Christian and I watched him change into a person who people wanted to be around. "

He believes that his rise through the Marine Corps was by the hand of God and in such God worked in his understanding of command. Civilians see the advertizing, 'The few, the proud, the Marines and understand the Marines are tough. Giving a man a second, even a third chance may not seem to fit the image of a select few but it does fit with the philosophy of a man seeing a grunt's value beyond his scars and the jut of his chin.

Offutt earned a final promotion to Sgt. Major before leaving the Corps in 2016. To obtain a promotion, a Marine studies the duties of each office, works to live by the Principles of the Marine Corps and appears before a Board of Officers who carefully examine his record of service. If a man is turned down for promotion twice, he understands that he will be leaving the Marine Corps.

The rank of Sergeant Major is the ultimate promotion for an enlisted man. A Sgt. Major is responsible for between 800-900 men and works with the Battalion Commander. He serves as part of a commanding triune, consisting of a Battalion's Commanding Officer, the Executive Officer and his role as the Sergeant Major. His men know he has proven his ability, his faithfulness and his service. He has earned his rank by merit, not by favoritism. He is responsible for all things pertaining to the enlisted men under his command. He is ready to give his Commander the history on each man in the Company. He learns these details by interviewing the men and reviewing their records before they meet their Commander when seeking a promotion or in an inquiry.

Sergeant Major Luke Converse served as Charlie Company's 1st Sergeant as the 1/5 enter Iraq. He later left the 1/5, earning the rank of Sergeant Major, a non commissioned officer in command of a Battalion. He describes the rank of Sergeant Major as "walking in circles." As the men of the 1/5 trained at Camp Pendleton, he moved from one company to the next, squad by squad, checking on their activities and their performance in their assignments. Returning to his office, he faced the drudgery of paperwork as he reviewed orders and the performance reviews that crossed his desk. Converse served in OIF 1, the invasion of Iraq and in OIF 2 in Fallujah before being transferred to another Battalion and deployment to Afghanistan in 2009.

Unlike the younger men under his command, he had come through the first Gulf War in 1991. He had begun his career in the Marine Corps as part of a three-man communication squad with the Joint Communication Unit that transmitted activity for all four branches of the Armed Forces. The Squad was assigned to set up a communications link in Kuwait, sending real time communications directly to Central Command and the Pentagon. This men were 'the eyes on the ground,' reporting what was happening as the Iraqis invaded Kuwait.

Along with the other two men in his team, he was captured by the Iraqis and transported to Bagdad. Shortly before being discovered, they destroyed their radios, equipment and documents along with their military identification. Converse and the other two team members blended into the diplomatic corps, allowing the Iraqis to believe they were civilians. To soften his appearance as one accustomed to military discipline, he grew a beard and wore civilian clothes.

Hussein knew that the Coalition would invade Kuwait to overturn his occupation of Kuwait and he found a use for his captives, stationing two or three at each critical site. Converse was placed at the Ministry of Defense. The Iraqis never learned that he was an active member of the US Military. Over four months, the days ticked by slowly with little to occupy his time. He was enveloped by a sense of helplessness. AS he trained his men to invade Iraq, He had intently studied the maps of Iraq, the terrain, the location of cities and town and specifically the water sources. With lots of time to think, he mentally reviewed the maps should he be able to escape and run for the border.

During this time, his relationship with God took on new relevance as he understood that he had no power to change the circumstances in which he was kept. The day came when the Americans were released as Hussein hoped to stave off the invasion. He was not successful as the Coalition had determined that Hussein would not bow to international pressure and destroy his chemical and biological weapons.

Held captive for four and half months, Converse was released on December 10. He returned home as American planes began pounding Bagdad. Forty-six days later he returned to Iraq as part of the invasion to liberate Kuwait. His Battalion was assigned to enter the northern reaches of Iraq among the Kurdish people The Kurds hated Hussein and the Baathist Party. They had suffered abuse for years which culminated in a chemical attack on one town, killing hundreds of men, women and children. The Americans were treated as liberators.

In 2003, 1st Sergeant Converse returned to Iraq with the 1/5. Unlike the young men around him, he understood the nature of what they would encounter. Now he was married, his wife expecting their second child - a baby girl. His commanding officer, General Fred Padilla, respected his experience with both the Marines and the Iraqis. On April 9, Converse would experience combat unlike anything from his experience during the first invasion of Iraq. And, he would live to see his wife and daughter when he returned.

He found after retiring from the Marine Corps that he personally had little to offer. The rigors of serving his country had stripped him of everything but the Corps. Months later, post traumatic stress began to set in.

"When I got out of the Marines my world fell apart," he recalls. "My mind kept going back to the time when I was a captive in Iraq. I had lost men under my control. You can have the best officers in command and men still die. I felt that loss. It was a chink in my armor. I couldn't let it go and allow myself to enjoy life because I held myself guilty for losing men. My world fell apart. I was miserable and angry. All my training and experience made me an incredible Marine but I was a failure as a human being.

"When my marriage went downhill, I began to understand that I had issues. Where you're around active duty personnel, everyone is the same as you. But now, I was around civilians. I found I was thinking about how I would kill myself and this scared me. I called the Veterans Administration and left a message.

"A counselor called me back. She was completely booked up but she said she could give me 10 minutes. I talked for five minutes and then she talk for five minutes. She asked me to write the details of one incident in Iraq. I filled ten pages. When I met with her, we talked about what I had written. She gave me more questions. In following sessions, he would write about an experience and they would talk about the responsibility he felt for what had happened.

In time he came to understand that he could not have controlled the actions of every man involved in the combat. He could not control when and where an insurgent chose to place an IED. He could not solely carry the responsibility for the injuries, the deaths and destruction that he witnessed as the 1/5 fought their way into Bagdad and later in Fallujah. The burden of trauma and guilt lifted. His mind began to heal.

After his first marriage ended, he remarried and is enjoying his role as father of three grown daughters. He is eager to tell you how proud he is of both girls in their accomplishments. Both his wife and eldest daughter work as nurses, serving in their own right.

After retiring from the Marine Corps, Converse began working with Security Services for the hospital in his small town. He sees his position as much more than standing watch over the hospital. He shuttles patients and fills requests when he has time from his officials duties. He perceives his role as being one of service, willing to serve where needed.

Converse has a story to tell those veterans who seem lost in the past. In a sense, the ultimate command is to speak the truth of what is possible. Converse no longer sees combat as a happenstance moment in time. He will tell you that God was in the events that occurred, preparing him and other Marines for what was to come. He still thinks of himself as an example for other Marines, as well as other men and women who seek relief from what troubles them. His story is one of a man who successfully made the transition back to civilian life.

In thinking of those lost and those who struggled to make the transition to civilian life I spoke with two corpsmen who had served the Marines of the 1/5. The Corpsmen from the US Navy are an integral element in the advance of a Marine Battalion. They run alongside the Marines as they come under fire and wait their moment to render aid. Al Alcala and Ken Huso stepped forward to describe their care for the men of 1/5 injured or lost in combat.

Chapter 9

Corpsman Up!

Chad Shevlin tore open the envelope from his wife, pulling out a photo. Staring at the image, a wave of resolve swept over him. He had a daughter!

"I will return to see my baby girl," he thought. He wrote a note, telling his wife how proud he was of her and how beautiful their little girl appeared in her first photo shoot. Then, he closed his eyes to sleep. Not more than an hour later, his 1st Sgt brushed the tent flap aside.

"Everybody up! We're going now!" An adrenalin surge swept over the men as they pried the sand from there eyes and struggled to their feet. Shevlin pulled his camo blouse over his head and stuffed personal items into his pack as he kept an eye on the twelve men in his squad. As squad leader, he would take them through the crossing into Iraq and Iraqi defenses. An hour later, they stepped into the amphibious assault vehicle, the door slammed shut and the tracks turned north.

Over 20 days, in April 2003, the Marines of the 1/5 swept north, turning aside all resistance. On April 10, their orders were to enter central Bagdad and take Saddam's Almilyah Palace. They halted briefly as humvees with non-essential personnel pulled to the side of the road. Onward, the tracks rumbled along the asphalt. The officers had been briefed. Over a thousand Fedayeen fighters from throughout the Middle East were awaiting them as they entered the city. All was quiet as the lead vehicles of Alpha Company passed through a ring of apartment buildings on the outskirts. Too quiet.

At an unseen signal, chaos erupted as gunfire hit the convoy from all sides. The men surged to their feet to meet the fire descending through the darkness. The gunners atop their AAV's firing, the men in the back shooting from open hatches, all exposed to the bullets and rocket-propelled grenades. Alpha Company took a wrong turn and doubled back on the convoy, searching for the route to the palace. Bravo led a long line of AAV's and humvees engulfed in a hail of bullets, followed by RPGs. Men were taking shrapnel as the RPGs exploded into vehicles. Their return fire blasted through windows and doors, knocking chips in brick and mortar. The radio spewed calls of "Corpsman Up! I need a doc!" Imagine, bullets flying through the vehicles as corpsmen braced, ready to jump, running to the wounded. Then the call, "White's hit!"

Gunny Bohr flicked his radio mic to call for a medic as he fired his M-16 toward the flashes of artillery from the buildings. In the next moment, he sank forward onto the dash. Gone! A bullet had pierced his forehead. The cry echoed over the radio.

"Gunny Bohr is down." No time to stop and check, no time to grieve. The vehicles surged forward, M-16's and .50 caliber weapons taking fire! Alpha Company followed a road alongside the Saddam canal, reaching the 15-foot walls of the Palace.

The guns fell silent for Bravo as they raced toward a traffic circle enroute to the Palace. The quiet was ominous after the chaos of the previous hours. A lone gunman stepped forward to fire an RPG toward the armored vehicles. The RPG blew past the air watch panel, striking Corporal Dawani in the hand before shattering the jaw of Corporal Chad Shevlin. A piece of shrapnel entered Private Moore's wrist severing an artery. Shevlin fired toward the Fedayeen who had pulled the trigger, killing him and then sank back into his seat. Blood flowed across his face from wounds to his jaw and throat. Glancing over, one of his men screamed, "Shevlin's dead!"

Shevlin grabbed the Marine, pulling him forward to thrust an upraised thumb in his face. One of the Marines pulled out a first aid kit to wrap the wounds and slow the bleeding. Then, he turned and fired again as they fought to reach the gate in the wall. Inside the compound, Shevlin crawled from the AAV and walked over to an improvised medical station where he was strapped to a gurney to prepare for evacuation.

He was choking, blood filling his esophagus, stopping the air from moving into his lungs. As an officer passed, he tapped the man's hand and signaled for something to write as he was unable to speak. The officer helped him sit up and belted him into a seat before lifting off.

Imagine working under such conditions with men who are bleeding severely or burned over much of their bodies. They look to you to save their lives as you tend to the wounds and prepare to move out, if transport is possible. Navy Corpsmen are critical in the actions of a Platoon in combat. They serve on the front line. As the men move forward or crouch in a defensive position, with bullets and rocket-propelled grenades flying, the cry for a corpsman rivets their attention. The corpsman may hesitate as he waits for a break in the firing or for a Marine to cover him as he races toward a wounded man. No platoon wants to see their corpsman down. He is their life line.

Nicholas Huso enlisted in the US Navy in June 2001. Four months later on September 11, he chose the option to work with a Marine Platoon. Like every new recruit, a corpsman first passes through boot camp before moving on to A-school and his initial training as a Field Medical Technician. Huso was given a choice of duty, either Hawaii or Camp Pendleton in southern California.

California? He had never been there. He thought, why not? Huso's first duty station was with the 1/5. He flew into Kuwait in February with the men of the 1/5, assigned to Charlie company.

The training for corpsman has evolved over the years with each conflict our nation has fought, both domestic and on foreign shores. Those wounded in the Civil War and the Spanish American War had little chance of survival. Gangrene, disease and inadequate care took their toll along with devastating injuries. World War I was not much of an improvement even as medical practitioners began to understand that the sooner a patient was treated, the better his chances of surviving the shock that followed. If a man did not succumb to his wounds, there was still a chance an infection would follow and antibiotics had not yet been developed.

Following WWI in 1928, antibacterial qualities in fungal growth were identified and penicillin was developed to fight infection. Our weapons of war also evolved. In World War II, 70 percent of battlefield injuries resulted in death due to a lack of medical treatment on site. When our nation entered the conflict in Vietnam, large numbers of troops were wounded but the number of those who died from their wounds dropped proportionately due to quick evacuation to medical stations for initial treatment. With the conflict in Vietnam, the Navy developed a program to train their corpsman beyond basic first aid, bringing them into the hospital trauma wards to learn how to care for the injured before the corpsman was assigned to a duty station in a war zone. A corpsman understands that he is not a doctor or surgeon. He is to begin advanced first aid and transport as quickly as possible. Corpsmen do not carry a rifle. They are armed with only a handgun for self protection at close quarters.

When a corpsman arrives in the chaos of battle, the Marines feel a sense of relief. The corpsman brings calm authority to the wounded. Though he knows these men, he sets aside an emotional response to remain professional in his assessment. Initally a medic would concentrate on the A-B-Cs of care. Under combat conditions, a corpsman first treats hemorrhaging. After stemming the bleeding, he quickly moves through triage, assessing 11 indicators that emphasize the A-B-C of care. A for airway - is it open? B for breathing - can the corpsman sense air passing into the lungs? And C for Cardio - does the man have a pulse? With severe bleeding to an extremity a corpsman quickly applies a tourniquet. A wound to the torso is packed with absorbent dressings. All of this should occur within two minutes so that the wounded, if able, can return fire along with his squad members. After the initial combat subsides, the corpsman precedes with the traditional standard of care as he assess how he can transport his patient to a medical station.

Each corpsmen wears a vest with multiple pockets containing the supplies he needs, from dressings and hardware to salves, quick clot powder and antiseptic spray. Many of the injuries for the 1/5 were from gun shots and improvised explosive devices that sent shrapnel into the bodies of the men. For burns, silver sulfide salve with a light gauze covering helps protect the burns during transport. The corpsman carries morphine to counter pain in severe injuries and comfort care. Time is essential as the corpsman works to quickly transport the wounded to advanced care.

Other injuries included impacts to the head, some severe, others were concussive. Vehicle accidents were common, whether in traffic or as the result of explosive ordinance. When not on patrol with his platoon, the corpsman may attend to basic medical treatment of his men and minor surgery.

Traumatic brain injuries (TBI) are not so easy to recognize and the Marines monitor each other's behavior for the signs of serious concussion. Nearly every man of 1/5 received concussions from the IEDs and other explosions. TBI is not always easy to identify when it first occurs and requires a long recovery as the brain attempts to isolate the damage and built new neural pathways.

Once the wounded man has been transferred to a medical station by humvee, the doctor makes an assessment as to whether the patient should be treated on site or evacuated to a ship or to Germany.

Marines take pride in being the toughest of the four branches of service. Many who were wounded were less concerned about the damage to their bodies and more about the men that remained in their unit. Marines regard their mission as the highest priority. In evaluating an injury, the corpsmen is evaluating who can be treated and returned to the front line and who should be sent out for advanced care.

On April 20, 2005, a squad from 1st Platoon, Alpha Company was caught by an IED as their humvee passed the Government center. Two men were killed, two others injured. Michael Tager, one of the wounded had a large piece of shrapnel embedded in his buttock. His concern was not for his own injury but for the badly injured Marine next to him. Across from him a second Marine lay in a fetal position on the floor in a puddle of blood. He assisted in removing the injured Marine to a transport even as his sergeant ordered him to join the remaining Marines in their raid on a suspected al-Queda stash of weapons.

Other Marines removed the shrapnel and Tager joined the surge toward their target. Only after the mission was completed, did Tager report to the corpsman for treatment. Despite being wounded and in pain, he had followed orders to complete their mission rather than inform his sergeant of his injury. After being transported to Camp Ramadi, he was examined by a corpsman. Pulling down his pants, Michael revealed a pair of boxers with smiley faces.

"Michael, I'm going to give you a tetanus shot."

"Doc, stick the needle right in the middle of the smiley face!"

"Your shorts are not sterile!"

"I don't care. Just stick it right in the middle of the smiley face!"

The Marines around him began to laugh, giving way to humor in their grief over those gone. As he climbed on to the transport, one yelled after him that this would be his ticket home.

Two weeks later, as two of the Marines from his squad moved equipment, a voice with a slow southern drawl rang out.

"Hey, Marines!"

No, it couldn't be! The voice with a slow southern drawl reminded them of Tager, as Michael, alive and well, crawled from the back of a truck the men stared at him.

"What are you doing here?"

"I snuck out. I couldn't let you guys down."

Tager was not the only man to make that run back to his company without being officially released from care. One of the biggest struggles for those evacuated is their sense of leaving their squad behind before the job is done.

The corpsmen build a special bond with the Marines in their squads. They understand that their training is a life line for the wounded men.

"You build a bond with these guys," he says. "Using your special skill set strengthens that bond. You know there could be a negative outcome when you're treating a severely injured man but I'm there to do everything in my power to save someone, to patch them up. The camaraderie you share in going through combat continues to this day. What I did in Iraq reinforced what I believed about giving the best care possible."

Twenty years later, the bond remains. Marines and Navy corpsmen easily resume where they left off. The hardest part for any corpsman was losing a man he could not save.

"That camaraderie was built in a group of about 12 guys. We built this brotherhood and that kept me in a stable place.'

It often took days for Huso to fully process the death of a Marine. Huso remembers one surgeon seeking him out after leaving the OR.

"Nick, there was nothing you could have done to save him. There was no intervention in the time this happened that could have saved him. You did the best you could for him."

This provided some measure of comfort but Huso found he still needed time to find an inner peace after leaving Iraq. Along with talking about his experiences, he understood that the sacrifices of the men he treated benefitted a lot of people.

"Their sacrifice allowed me to move forward. I was motivated to stay in the medical field, moving into radiology. It no longer hurts like it did. I still grief but in a different way."

When the tour of duty ends, the corpsmen may choose several avenues to pursuing further training. Some move into dentistry, others toward radiology or advanced nursing in a clinical setting. When Huso left the US Navy, he had risen to Master Chief, marking a career dedicated to service and the sacrifice of the men he had stood alongside.

Like Huso, corpsmen who remain in the Navy may choose to advance their education. Moving beyond a Field Service Technician position is the Surface Force Independent Duty corpsman (IDC). He is the primary medical officer aboard ships and in isolated duty stations. He works independently of a doctor on site performing diagnostic tests, nursing, minor surgeries and basic clinical labs. He may teach Field Corpsmen and review their work, assisting them in their careers.

The Special Amphibious Reconnaissance Corpsman may be the most challenging in terms of risk and hardship for they serve with the Seals, the Naval Special Op Force. The training is similar to a paramedic but the corpsman will be working in extreme conditions in covert operations. They must be able to work independently with a degree of uncertainty in transporting their patients to advanced medical care.

After leaving the 1/5, Al Alcala-Esparza chose to train as an IDC. Alcala is a native of Mexico and entered this country on a tourist visa at age 19. He delayed his return to Mexico without notifying 'La Migra.' He had been drawn to the culture of the United States and learned English by watching American programing on television as a boy. He liked what he saw in the culture and lives of young men like him.

After moving to the United States, Al realized that he wanted to serve in the Armed Forces but when he visited a recruiting office, he was informed that he must first become a permanent resident and then apply to become a legal citizen. This is a prolonged process and years later, Al was able to complete his enlistment. He selected a career as a corpsman. Despite retaining a thick accent, he enjoyed seeing patients and taking their medical history.

In time, after he left Iraq, the doctors began to complain that they could not understand the orders he had written on each man's chart. He had gone to war and tended the wounded as the Marines fought their way cross Iraq. Now, the combat had taken its toll. Too many explosions, to much pain and blood. All of it affects the neurological pathways in the brain and at first the corpsman may not realize that his brain has been damaged. The concussions from the explosions had given him a traumatic brain injury. Noxious substances and trauma had attacked his health. Alcala was informed that he would be retiring from the Navy. He was no longer able to fully function as an IDC. The neurological damage from combat ended his medical career.

I watched as he sat, waiting for a ride to the top of 1st Sergeant's Hill during the 2024 reunion. His chin cupped in his hands, elbows resting on his knees, his eyes staring at the ground. I wondered if regret seeped through his memories as he spoke with the men he once treated for their injuries. He still keeps in touch with many of the men of the 1/5 and attends the reunions. He reaches out to the families who have lost a son in Iraq and serves as one of the liaisons to the 1/5.

One evening Doc Al sent a text, suggesting I talk to his former ambulance driver. When the 1/5 first entered Bagdad, some of the corpsmen were driving their own vehicles but they were not trained in the coded commands given by the Marine Corps Convoy Commander. The convoy came under fire and the Convoy Commander ordered the convoy to push through the area, yet one ambulance ground to a halt without warning. The dust and smoke from the long convoy, coupled with lights-out driving, made it difficult to see the vehicle ahead.

PFC Tony Reed was behind the wheel of a seven-ton truck loaded with ammunition and explosives, driving and navigating with night vision goggles in a convoy without lights. Visibility was poor, and out of the gloom he realized he was bearing down on a stationary ambulance with Navy Corpsman at the rear of the vehicle. He was too close to the ambulance to stop. He slammed on his brakes and shut his eyes. With the impact, he was certain he had just killed a corpsman. When he dared to look, he realized the ambulance had pulled away, doors closed, leaving him with a stalled truck, but he found no bodies. The Navy corpsmen had jumped into the ambulance a moment before impact and rejoined the convoy.

The crash and stalled truck caused the convoy to split in half. PFC Reed and two fellow Marines retrieved an extremely heavy tow bar, retracing their route past 15 trucks to the rear of the convoy. Together they lifted the tow bar and ran back to the 7-ton truck. Using the tow bar, they connected to a second truck and were towed to a safer location where they rejoined the back half of the convoy. The Marines were ordered to remain in this location due to a high risk that they would come under assault if they were to mobilize toward joining the rest of the convoy. As they waited for two days, the men repaired the truck. Following that incident, the Commander decided that Marines from the Transportation Company would drive the ambulances and the Navy Corpsmen would be solely responsible for treating the wounded.

As 1/5 deployed to Fallujah the following year, newly promoted LCpl Reed was assigned to drive the ambulance for Doc Al. The men came under extreme duress again the morning 1/5 began their assault into Fallujah. Understanding their duty to tend the injured, they entered the city without an escort. The insurgents were in control of Fallujah and the men were in extreme risk of being sacrificed to their mission. Just weeks earlier, four contractors had been caught in the city and executed. Two of the contractors were strung from a prominent bridge in Fallujah.

LCpl Reed's ambulance came under fire while driving through the city, searching for the squad who had initiated the call. Later, they would be told never to enter the city again but to remain on the outskirts. The Marines in the heart of Fallujah would bring out their wounded for medical treatment for the sake of efficiency and safety for all.

I recall reading the horrific stories of ambulance drivers dodging potholes and deep, muddy ruts in the French countryside as bullets streamed past and cannon shells soared overhead during World War I. While US forces are now better equipped to handle the terrain of enemy territory, the experience for these two men seems to echo what ambulance drivers suffered in previous conflicts. The trauma for these men is real.

Their job was to drive the wounded. At times, they offered emotional support to badly wounded men. LCpl Reed recalls one injury with a bullet passing through the artery of a fellow Marine's neck. The Marine was suffocating, choking on the blood flowing down his esophagus. Anger radiated from the Marine as he fought to breathe, eventually losing the struggle. LCpl Reed understood the anger was not directed at him personally but he grieved over his inability to medically intervene.

When his four years were up, he had advanced to the rank of Corporal. He did not re-enlist but chose to return home. He found life at home was not the same. The transition from Cpl Reed to Tony, as his family called him since boyhood, felt like it was stripping him of the identity he developed while serving the Marine Corps.

Raised in a family who followed the Protestant faith, Tony, like many others struggled to understand how God could allow what he had experienced to happen and he lost all faith. His marriage disintegrated and he began drinking heavily.

He was no longer the carefree teenager who left home to serve as a Marine. He felt lost, without direction or a sense of self-worth. Entering college, he found the academic culture not welcoming to a combat Marine. His family did not understand what had become of him and he withdrew, sinking into anger and depression.

In 2008, he began dating a woman he had known in high school. As their relationship became serious, she encouraged him to begin climbing out of the pit he had fallen into with the trauma from Iraq. They married in 2010 but Reed continued to suffer from survivor's guilt and loss of identity. He did not seek treatment until suicidal thoughts began to plague him.

In 2016, he suffered a full meltdown after being bitten by his cat, seemingly a small thing. The cat was defending itself when Tony became angry over damage the animal had caused. His cat served as a therapy animal, often grounding and calming him through his anger, flashbacks and episodes of PTSD. The bite required medical attention. Due to the dormant memories the incident awaken, Tony spiraled into despair and his wife drove him to the Emergency Room where the couple discussed the options for inpatient psychiatric treatment and other levels of mental health-care.

In seeking mental health care at the VA, he underwent prolonged exposure therapy, telling and re-telling his story of the Marine dying. With each re-telling, more details were revealed until with the final episode he acted out the scene, complete with accents for each man. The process was exhausting but Tony was able to let go of the emotional impact. He felt as if a weight had lifted on his shoulders. He was released to live again.

On the anniversary of his most challenging moment, a slight depressive consciousness creeps over him and he takes a few days to settle and move past the memories. He now believes that living a peaceful life is the best way he can honor those who's lives were cut short. He and his wife now have two children. As time allows, he develops his artistic talent, creating paintings of how he felt in Iraq as well as portraits of those he loves.

The two other men I've mentioned in this chapter, Michael Tager and Chad Shevlin would go on to successful careers after being discharged. Tager began a career dealing in fire arms. In 2024, he opened a store front near Houston, Texas.

Chad Shevlin aspired to a career in law enforcement. He overcame a home life with a violent, alcoholic father and enlisted in the Marines only to be severely wounded. He underwent multiple surgeries to repair his jaw with extensive dental work. Plastic surgeries eased the wounds to his face and neck. His vision is impaired in one eye. The injuries did not end his ambition to serve the public in law enforcement. He became a police officer in his home town of Raymond, New Hampshire and reached the rank of Captain before being hired as Chief of Police for the neighboring town of Candia in 2024.

Each of these men have met irreversible damage with a smile and new goals as they overcome life-altering injuries. Each has struggled with the nightmares and neurological triggers that dangerously shred their emotions. They know well the devastation of depression but this did not stop them from setting new goals and striving to accomplish them.

With his recovery, Chad Shevlin began to train for his first marathon. When he speaks in schools and social organizations, he reminds his listeners that we do not walk out the door one day to spontaneously run a marathon. We start with the first step. We run a mile, increasing the distance as our strength and endurance improve. So it is with healing, one step at a time.

One corpsman talking of the years after Iraq, said, "Some guys get stuck in the moment. They let themselves go. You've got to use the experience as a stepping stone to do greater things."

Chapter 10

Families of the Fallen

Marines are trained to understand that tough challenges are part of their service. When these men step back into civilian life, how do they go from the blast of sudden gun fire and explosions to changing diapers and an 8 to 5 job? For many of the men returning from Iraq and Afghanistan, that challenge is not one they are prepared to meet. They have grown accustomed to the rush of adrenalin. They feel alive in emerging triumphant over risk and danger. For many, in the quiet of the home ground, anger and depression become their companions. They may believe that seeking help is admitting a weakness for a man who has been trained to rise above the fray. As a result, our combat-worn veterans look for alternative means to mute the pain.

Government statistics showed the number of U.S. military veterans receiving medical care from the VA reached 6.2 million, nearly double from 3.2 million veterans at the end of fiscal year 2018. Veterans have seen a spike in urinary, prostate, liver and blood cancers over the last 20 years. The Veterans Administration credits an accounting error for the steep rise. Liver and pancreatic cancer also increased sharply.

Veterans have noted an increase in neurological conditions such as Parkinsons. The Veterans Administration instituted a rule giving veterans just seven years to seek medical help after leaving Iraq and Afghanistan. Consequently, not all veterans with these illnesses are enrolled in the official tally.

Medical conditions are not the only cause of early demise among veterans. Poor decision-making and a need to mute the trauma follow the veterans of Iraq and Afghanistan. One Marine said, "When I put my life on the line, I feel alive again."

On October 26, 2023, an article in the USNI News quoted the Pentagon, saying, "Marines See Highest Suicide Rate Since 2011, Navy Since 2019." According to Pentagon statistics in 2011, 15 out of 100,000 veterans committed suicide. In 2022, the number increased to 34.9 deaths per 100,000 veterans. The Marines Corps has recorded the highest number of suicides proportionally out of the four branches of the Armed Forces. Many come to the point where death seems preferable to their struggle with depression and post traumatic stress.

The Marines of the 1/5 count these men, the ones who have died after returning home, as casualties of their deployments to Iraq and Afghanistan. They understand that the injuries and trauma these men endure did not end when they received the official discharge papers. The battle continued, carried to the home front. Not all survived.

When veterans carry the pain and loss from battle into their homes, the fight to recover their lives shifts from the military to their families. Civilians often do not understand all this transfer brings to the wives, the children and the parents of the Marine. Some of the men had been through painful separations from their partners and families.

The number of men from the 1/5 dying prematurely became the impetus for Major General Fred Padilla as he called for a reunion on the 20th anniversary of the entrance into Iraq. If the men of the 1/5 came together once again, he hoped they would find a sense of belonging and receive encouragement from talking with others. He hoped recovering a sense of camaraderie would draw men out of their struggle with post traumatic syndrome and painful memories?

To better understand the struggle of these men, consider the stories of four Marines who have succumbed after emerging from combat. First, two important points. The Marines consider the men who have died post Iraq as casualties of the combat they experienced when deployed. These men are honored just as any man who died in Iraq or Afghanistan. They were as tough as any Marine. Yet, when they returned, the battle did not end.

With both the mothers of the fallen as well as those who lost their sons in Iraq, many would quietly tell you that they do not want to be known as the woman who lost her son! We are proud of our sons and the choices they have made. We stand behind them and love them in the midst of our memories and we ask that others respect this.

Weeks after Enrique Ramos died to the ravages of cancer, his wife, Bianca, swept through the garage, clearing out the items that had accumulated over the previous ten years. She noticed a small book, half hidden. Pulling it free, she recognized her husband's handwriting flowing across the pages. His words are of Enrique's struggle against depression and anger, even in his love for his family.

Enrique Ramos enlisted in 2003, and was assigned to 1/5 Bravo. Bianca had known him since age 14 and became his wife three years later. He left the Marines, honorably discharged in 2011. After his discharge, he complained of pain in his knees and shoulders. The doctors repeatedly injected cortisone into the joints, trying to reduce the inflammation. In early 2017, he learned he had developed stage 4 colon cancer. Enrique died Sept. 16, 2017.

As Bianca sat in the hallway of the hospital after he had passed, she wondered what she was going to do. She was in her final trimester of pregnancy, expecting a baby boy. How was she going to pay the mortgage on their home? How was she going to pay the utilities and other bills? And how was she going to raise their three children as a single mom?

Her insurance agent came down the hall and she looked at him, asking, "What am I going to do?"

"Remember, we planned for this," he reminded her. "Months ago, Enrique bought an insurance policy. The money will give you time to raise your children in their home." "I felt as if a heavy weight had been lifted off my shoulders!" Bianca recalls.

Historically, the Hispanic population does not purchase insurance. They choose to put their money in the bank rather than purchase an insurance policy. During the years they were married, Enrique wanted Bianca to raise their children at home.

"He believed no one would watch our children like a mom," she said.

This could be challenging as California's economy boomed and the cost of living boomed. Now he was gone and she had the future to consider. Rather than pay off her mortgage, Bianca chose to invest the money she received from their insurance policy. This gave her a small income. She studied to obtain her license as a financial advisor and insurance agent. Her son, born just weeks after Enrique's death, slept in his baby carrier beside her.

"As an agent I can set my own hours," she says. "I could make

---

A Journal Entry written by Enrique Ramos before his death from cancer.

The reason why this happened is because we made mistakes. Mistakes that lost us a life. We went into a very hostile area that we knew was dangerous and should have not gone into without some reconnaissance. We went in and came out one body short. Until this day deep inside of me, I know I should have been the one not making it out that day. It's a feeling that doesn't leave me. It follows me everywhere. I was the target they would've hit. I tried to see it as him saving my life and it's difficult for me to accept that he's gone. He had his family to come home too.

Now I have my two daughters and it causes pain because he didn't come back to his two daughters, which were the same age as my daughters are today. The day of memorial I was ashamed to look them in the eye because their daddy did not come home with us. They were too young to understand that he wasn't coming back. That he was dead. I imagine my girls receiving this news. It is painful to think of my girls calling someone else daddy and my wife being with someone else. I lose my damn mind. I will never be the same person I was. It is impossible. I've been through some shit and have caused so much harm and damage that I can't feel clean or at peace. I made it home, but everything came back with me. The thoughts, the images, the feelings, the hatred, the pain injuries, everything. It has made my life difficult and very miserable at times.

The class I've been to at the VA has made me realize that my quality of life sucks. The fatigue and lack of sleep, paranoia and

---

appointments when I was available while the children were in school. This worked well for me as I earned additional income."

Finding an occupation to support her family was not her only hurdle. Enrique's family chose to claim ownership of her home, a house she had purchased with Enrique. They claimed the furniture as well. Without a will, Enrique's estate would be settled through probate before a judge. She recalls standing in the courtroom as he handed down the decision.

"He must have looked at me as a single mom with no husband, no income. He looked at my three children before awarding the

> fear have made my life very difficult. It is affecting my family and my marriage. I'm bringing them down the hole with me. I don't want to see anyone get hurt again. It is painful for me. I'm afraid to see that again. I am the only one that can keep me and my family safe. I don't trust anyone else to do it. Everyone is just looking out for themselves. That's why I feel safe staying home. Being outside friendly lines is always dangerous. I will use what I have, been taught to do in order to keep my family safe no matter what I have to do to keep it that way. I'm already lost cause. I won't trust anyone anymore. I have trusted people and that outcome has been the same. A stab in the back. Everyone just cares about themselves. I will just keep everything to myself. It is my problem anyways. No one else cares, and they will never understand. I used to think I had people that I could trust and I could open up to, but that ended quickly. It irritates me that they go telling people that I'm the problem. They have no fucking idea what I'm going through or what I feel. I feel rejected and all because of what I'm going through. They see me different from everyone else. I don't feel close to anyone anymore. The connection I had with people is gone and it doesn't bother me. I am my own team. I've been betrayed by those I never thought would do such a thing and it has caused anger and pain. It has even affected the relationship I had with my wife as well. I feel numb and cold inside. I fight these feelings everyday. Being alone and away from people has become me now. If I trust again, I will only bring bad things. I can't deal with more negative shit. All I can do is deal with myself, my life one day at a time. As difficult as it gets, I am alone with my fight.

house and furniture to Bianca.

"I was so grateful. I still had a home! she says."

The Veterans Administration also denied benefits to her after Enrique's death and she chose to sue. A single mom suing the Veterans Administration of the Federal government? Was she intimidated? What if she lost and had to pay the attorney's fees? She laughed and agreed that it was a bit intimidating. Enrique had served almost 10 years, part of that service in combat in Iraq.

"I found an attorney who would represent me without money up front. We won the case and in 2024, I was awarded his benefits."

The settlement came as her oldest daughter was about to enroll in college. She had fretted over how she would pay her daughter's tuition and expenses. The settlement was very timely. Her attorney thought she might have had a case for malpractice if she had acted sooner.

Having learned from her experience, Bianca has spoken to groups of other women and employees of Priamerica, her agency, about what she has learned. She hopes her story will inspire others toward wise financial planning.

Looking back on her husband's struggle with cancer and PTSD, she remembers the months before he died as hard, really hard. Words and events were triggers that caused him to react as if a light switch had been flipped.

"I took some classes at the VA to try to understand what was happening. I came to understand that his anger was geared toward the person he knew would never leave him. This was hard to accept.

"He would ask for a divorce, saying, I don't want to drag you down the hole with me. I would tell him I wasn't going to leave him and that I loved him," she says. "Enrique sought counseling but it didn't seem to help the anxiety, the depression or the mental and physical pain. His depression and pain turned to anger and that anger was directed at me. I heard this from other women in the group as well. The counseling and treatments didn't seem to help. He felt as if he didn't receive the help he needed. Before he died, his anger got a lot worse. He was verbally abusive. No physical abuse, though at times I wondered if he might kill me.

"I might go to McDonalds with the kids to pick up food and I would call him just to make sure he was okay. As his care giver, I had no time for me, no relief to care for myself. Life revolved around him.

"After he died, it was a relief from his anger and the abuse. I found the small notebook in the garage and I was so sad. I just hurt for him as he was unable to find the help he needed. I've wanted to share his story and this is my opportunity to do so.

"Now I'm finding me and learning who I am. I love sharing my story as it impacts the lives of other people. I want to help others plan for the future."

As I said previously, the battle these men engage in shifts from the battlefield to the home front. Their families become their comrades in the fray. Bianca struggled against cancer as do many families. Two others would witness their sons emotional struggle, using alcohol to ease their inner conflict.

Roy (Bud) and Laurie Claar raised their family in rural Pennsylvania. Their home town is carpeted by mature trees, green lawns and historic buildings in the foothills of the Appalachian Mountains. As one of four children, Matt was the next to youngest, eager to follow his dad and big brother outdoors. Early on he showed a reckless streak, deciding to bungee-jump from the tree in the front yard. He cut a length of rope and tied it around his waist, then launched from a branch well above the ground. He had not measured the rope and found himself dangling from the limb, toes high above the ground. His brother cut him down, laughing at his predicament.

Matt wasn't all that enthusiastic about school, earning average grades. His preferred method of learning was hands-on. He longed to be outdoors and enjoyed hunting white-tail deer with his dad. In high school, he nearly severed a finger as he began dropping a hatchet in a contest to see how close he could come to his fingers without sustaining an injury during a camping trip with his cousin. He wasn't one to back down from a challenge.

His junior year of high school, he determined to enlist in the Marines, following the example of his grandfather and his uncle who served in the Navy and the Marines respectively. Bud refused to sign, saying Matt could wait till he was old enough to sign on his own. After turning 18, he enlisted with a buddy and was sent to Paris Island for basic training and then on to Camp LeJeune. He arrived at Camp Pendleton, assigned to 1/5 in 2003. He would be deployed twice, once each to Fallujah and Ramadi, both cities with fierce fighting in the Anbar Province of Iraq. He didn't talk much to his mom about his experiences.

He told her, "I love the Marines but I hate the politics," referring to the wrangling taking place in Washington, DC.

At the end of four years, he did not enlist, choosing to join his father in the family's roofing business. Over the next year, he engaged in what his mom called reckless behavior and drinking excessively. Laurie remembers one episode with Matt 'clearing the house' with a loaded pistol during a bout of post traumatic stress.

One evening while visiting a friend, another man entered the house, slamming the door behind him. Matt reacted and the confrontation grew into a fist fight with the visitor using a broken bottle as a weapon. Matt sustained severe cuts to his back, requiring 98 stitches. On another occasion, he was sleeping on the couch. Laurie thought he might be cold and covered him with a blanket. Startled, Matt sprang to his feet, ordering her to never do this again. Both incidents indicated Matt was struggling with post traumatic stress.

As trash accumulated, the family burned the refuse in a barrel. One afternoon Bud dumped gas on the trash and lit it but the fire was not consuming the refuse quickly. Matt threw more gas on the fire. The flames arched along the stream and ignited his sleeve, reaching from the wrist to his shoulder. Medical personnel who treated the burns sent him home but infection began to set in. He entered the VA, seeking treatment and medical personnel, in turn, sent him to the West Pennsylvania Burn Center where skin grafts were applied to the second and third degree burns.

Months later, Matt checked in for his first appointment with a counselor, hoping to find relief from the significant signs of PTSD. Like others, he had been using alcohol to self-medicate. Due to his drinking, he racked up two citations for driving under the influence and was placed on home confinement with an ankle monitor. He could go to work but he was required to come straight home.

Some days he played with his gun and talked about how it would feel to be shot. He went so far as to put on a vest and ask a friend to shoot him so that he could feel the impact. His friend refused his request.

In September 2008, the Claar's granddaughter became very ill and was taken by a medi-flight to Pittsburgh. The following Saturday, Bud and Laurie drove to Pittsburgh to be with their older son and his family. Matt stayed behind. Early in the morning hours of September 21, he pulled the trigger of his gun, taking his own life.

The reasons for Matt's decisions are complex. Neurologist understand that the frontal cortex of the brain does not finish maturing until around age 25. Matt had just reached that pivotal point. He had suffered two serious injuries in the two years prior to his death and experienced several incidents that triggered episodes of post traumatic stress.

Bud Claar speaks out on behalf of the veterans struggling with PTSD saying, "The Marines take these kids and give them all this training. When they come home, the Marines do a debriefing, take their combat boots, given them their sneakers and they're out. They need to take as much time to debrief and talk with other Marines as they did in training these men."

This is a valid point though the return for the investment in time and resources may be incalculable. The home confinement may have helped stabilize his daily routine and the counseling he received from the VA assisted his struggle as well. Ultimately, we do not know why Matt chose to surrender his life. This is true for many families as they struggle with the death of a child. The question of why a young life is released so early is not an easy one for parents to understand.

For Laurie and Bud, their faith plays an important role in coming to acceptance of what has happened. They have made some good decisions since Matt's death that help them work through the grief. Laurie works with another friend to collect funds to assist the families of veterans. They call the drive, Funds for Freedom, one year the fund was able to assist 21 families. She has reached out to another mom whose son had committed a serious crime, standing beside her through the legal proceedings for her son. After we experience the earliest stages of grief, it is often important to work toward an objective.

Like many mothers, Laurie wants her son to be remembered for how he displayed a tender side toward those he loved with a sense of humor. Along with honoring Matt at two weddings, three of Laurie and Bud's grandchildren have been named after Matthew.

"We made a choice. You never forget," Laurie told me. " We understood we could let grief destroy us or we could choose to accept what had happened and move on with our business, with our children and grandchildren." The couple have 10 grandchildren and five great grandchildren.

Many veterans use alcohol or drugs to mute the grief and pain they experience after returning from Iraq. On first thought, we may not consider excessive use of alcohol as a means of surrendering one's life but the family witnesses the slow deterioration alcohol brings and the seeming inability to remain sober.

Joshua Kegley deeply admired his father who had served with the Marine Corps during the years of the Vietnam conflict. As a young boy, Josh loved playing with his plastic army men, lining them along the edge of a table and shooting them off, one at a time. School wasn't all that important to Josh growing up. He loved being outdoors with his friends and his brother, Jason. After leaving the Marine Corps, his father, Tony, went to work for the railroad and was often gone for days at a time. Lee Ann, the boys' mom, chose to work outside the home, entering management with the McDonalds Corporation.

On September 9, 2001 Joshua enlisted in the Marine Corps and shipped off to boot camp, intent on making his father proud of his service to their country. Two days later, as Lee Ann worked out in the gym, she watched two planes fly into the World Trade Center on the television monitor.

"I knew in that moment that Josh would be involved in the response to the attack," she recalls.

After training at Camp Lejeune and Camp Geiger, Josh was assigned to the 1/5, based in Camp Pendleton in California. A year later, the men of the 1/5 began loading for the flight to Kuwait.

Arriving in a barren stretch of desert, they built a camp, trained and waited for the order to cross the border to challenge the armies of Saddam Hussein. The order to move forward came and Josh took his position as a machine gunner on an amphibious assault vehicle.

Six months into the deployment, Lee Ann received an email.

"Mom, I can't tell you when we will be home but please be ready to fly out on short notice to meet the Battalion when we return. Your face is the first one I want to see." Lee Ann packed her bag and checked the airline schedules.

Lee Ann was not the only one Josh was thinking of as he returned. In the seventh grade, he met Erica Howell, a lovely young woman with dark hair and a quiet demeanor. Josh and Erica dated through high school. On October 4, after their return, Josh called his mom to tell her this was the day he would marry Erica, his high school sweetheart. He told his parents he was ready to settle down. Five months later, Josh once again prepared to enter Iraq with the 1/5 as a sniper in the battle for Fallujah, a key city in the Sunni triangle, where Hussein received his most fervent support.

Later, Josh described long days and nights laying on the roof of an Iraqi home, surveying the territory below. The men at times did not have food or water. Leaving his sniper post to get a drink or to relieve himself in the bathroom was considered a privilege. On one particular day, Josh was forced to shoot a young girl wearing a red dress. For Josh, who loved children, this was beyond what he expected to encounter. The girl was carrying a bomb.

His sergeant reassured Josh, saying, "You saved a lot of lives out there today."

In describing the moment to his mother, Josh said, "I still shot her, mama."

He could not seem to move beyond the guilt of killing a young girl or other incidents that had occurred as the men fought for their lives and control of the city. When Josh's term of enlistment ended, he returned home and took a job with the County Sheriff's Department.

In time, Gunner, a baby boy, was born to Josh and Erica. The family attended church regularly and Josh sang with the team leading the worship service. He struggled to reconcile his actions in

Iraq with the faith he had received at an early age. He was haunted by nightmares, by loud noises, flashes of light, all symptoms of post traumatic stress. Due to the PTSD, he was given 80% disability.

After eight years with the Sheriff's Department he resigned and became a stay-at-home dad, doing many of the household chores while Erica supported the family financially. In 2016, a second son was born. As Talon grew, Josh took the young boy with him into the woods, riding a 4 x 4. He taught both his sons to observe life around them as well as the skills required in the forest.

Despite the routine of chores and times with his sons, Josh suffered multiple episodes of depression and flashbacks to his time in Iraq. He began seeking relief in alcohol. While under the influence, he had multiple episodes handling a gun when he seemed to contemplate ending his life. Gunner was witness to those moments. With each episode, Erica drove him to the Veterans Administration Clinic, seeking help. Josh received counseling and medical treatment. He would attempt to stop or at least limit how much he was drinking. After returning home, he would sink back into depression. This is a struggle that many households with veterans know well. Their families love these men but they feel helpless to know how to help them.

The Veterans Administration has created several programs to assist the veterans. Counseling and medication may also be obtained through private practice. This country owes our combat-tested warriors a debt but ultimately, we do not know enough about the long term neurological consequences of exposure on the battlefield to the explosions, to the killing and bloodshed.

Looking back to the aftermath of World War I and II, an older generation remembers the silence of the men who returned. Few were eager to tell of their experience as they fought the Axis forces. Some simply shrugged off any inquiry, others lapsed into silence, dominated by a long stare.

Josh, realizing his limitations, applied for 100% disability. He did not stop drinking. In time, one of the doctors in the Veterans Clinic leveled with him.

"Josh, the numbers from your blood test on your liver function are not good. Excessive alcohol is destroying your liver. If you don't stop drinking, you are going to die."

Men who battle post traumatic stress explain that alcohol does not allow them to forget what has happened. Rather, alcohol numbs the pain. In seeking relief, Josh was caught in the cycle of addiction. On March 12, 2020 he became delusional. His eyes had turned a deep yellow indicating his liver was starting to shut down. When Erica realized how desperate his condition had become, she called the emergency squad. At the clinic, the doctors intended to med-evac Josh by helicopter to a large hospital in Columbus but it quickly became apparent that his internal organs were shutting down. Erica made the difficult decision to remove Josh from life support. An ambulance carried him home where he passed away two days later.

There are days when Talon, a bright-eyed, red-headed boy, visits Lee Ann and asks to talk privately. They go up to her bedroom and he asks, "Mammy, why would God take a little boy's daddy from him."

Her heart breaks over his questions, his confusion as to why a loving God would permit this to happen. Each time her answer varies but she does not shy away from the truth. She reminds Talon that his daddy remains with him and his brother, Gunner. She tells him his mama still loves his daddy and that Josh is waiting for him in God's presence.

After Josh's death, Gunner began wearing the style of neck gaiter his dad had worn in Iraq. Throughout Covid, the gaiter covered the lower half of his face but as the threat diminished he continued to use the gaiter. Lee Ann thought he found a sense of security in wearing the band. When he arrived at the reunion, the gaiter covered the lower half of his face. As he relaxed around the Marines, he became more open and he lowered the gaiter.

Would the story have turned out differently if Josh had sought more help? This is a question many families ask. The resources are available. If a man will not step forward and ask for help, he may slip into despair and make choices that are harmful to himself and others.

This unwillingness to seek help in the struggle may have contributed to the death of Noah Pippin. The story of the Pippins' search for their son is one of his parents' love for their son, an account of unwillingness to end their search until they learned what took their beloved son from his family. The story also speaks to the reckless behavior some Marines indulge after leaving the Corps.

Mike and Rosalie Pippin met as they finished college and began professional lives. They brought three boys into the world, Noah, Josiah and Caleb, but Noah was to stand out as an active, independent young man. By the age of two, they found him strolling across the fiberglass roof of their porch after crawling out of a second floor window. By the time he was eight, he became determined to walk five miles home from school - alone. Rosalie remembers this as a different time when one could allow their children to wander without hovering over their every move. They made sure to drive the route with him, pointing out the hazards along busy roads before allowing him to venture onto the route alone. Noah arrived home safely.

He was diagnosed with attention-deficit. Though intelligent, he struggled in school. As he matured, he enjoyed reading and became an excellent writer. Noah was the first son to take an after school job, first mucking out stalls and caring for the horses. One stallion was particularly challenging but Noah was not afraid to enter his stall. He learned he could urge the horse to leave the stall with a pea shooter, flicking a pea against the horse's flanks.

Early on he made a decision to believe in God, making a profession of faith. He loved to sing hymns in church and regularly attended services. However, as he grew older he began to question the teaching he received, frequently arguing theological points. With a chuckle, Mike describes both Rosalie and himself as typical, manipulative Christian parents concerned for their son. As he reached his senior year, they insisted that he would attend either a week-long Christian seminar or a weekend Chrysalis retreat concentrating on the doctrines of the Christian faith. Noah chose the latter. He met a local pastor who was equipped to listen without being threatened or feeling the need to respond.

He started his studies at Central Michigan University, feeling comfortable in college. After two years, Noah left the university as his sense of responsibility did not find a home in the campus drug scene. He enlisted in the Marine Corps in 2003, listing his religious preference as atheist. As he prepared to ship out to boot camp, he told his parents they would not hear from him in a long time and he would not return to their home any time soon. After six weeks, they received a letter. Noah was homesick. He could not wait to come home, apologizing for his earlier comments.

As he moved into infantry training, Noah approved of the discipline and training. He found that disputes could be settled with in a boxing ring. Upon completion of training, he was sent to the 1/5 as they deployed to Fallujah.

Mike felt that Noah had found a home in the Marine Corps as serving in the Corps fit with Noah's desire to live with honor. He made several good friends and felt what he was accomplishing was worthwhile. Like many parents, they watched the news reports and prayed for their son's safety. When he did call, Rosalie would ask, "Noah, why don't you call us more often?"

Exasperated at her comments, he would say, "Aw mom, when I go by the communications center the line is long. I don't want to spend the short time I have waiting to call."

"We love you, Noah," she told him. "We want to hear from you and know that you're okay."

"Mom! If something happens to me, the authorities will contact you. If you don't hear anything, I'm fine!" That comment would haunt them years later after Noah disappeared.

Noah went on to serve a second deployment in Ramadi, then a third deployment before he was discharged in 2007. Rather than return to Michigan, he camped in his car at a rest stop for six months before being hired by the Los Angeles Police Department. He also enlisted in the California National Guar and was set to deploy to Afghanistan in December 2010.

In 2010, Noah called his dad, revealing he had been involved in a training incident with the police department. Due to his sense of honor, he did not feel that the incident could be resolved in a manner that satisfied him. He chose to resign from the department. Mike remembers feeling a sense of alarm. Something was not right with Noah. A friend in the LAPD later told the Pippins that Noah seemed very closed off in the months they lived together.

Noah did not want to ask for help when he faced a problem. He would figure out the problem on his own and this was to prove fatal in the mountains of Montana.

After resigning from the LAPD, Noah remained in Los Angeles, and began training for mountain warfare on his own. He told his parents he intended to give his possessions away. They suggested that he rent a moving van and bring his possessions home, storing the boxes in the basement. After he arrived, they enjoyed their visit but he was hesitant to tell them his plans. As their concern grew, Mike asked his son, "Noah, are you contemplating suicide?"

Noah turned halfway toward his dad but he did not meet his eyes. "No, dad, I would not do that!"

Nothing more was said. All too soon, he was making plans to return to California. He asked his parents for suggestions on locations to visit as he drove west and they suggested Glacier National Park. On August 25, 2010, as prepared to leave, he called a taxi. At the last minute, Rosalie snapped a photo as he talked to the taxi driver. He left all but a few belongings.

And then, there was silence.

On September 11, 2010, Mike and Rosalie received a phone call from Noah's National Guard Sergeant.

"Mr. and Mrs. Pippin, do you know where your son is? He didn't show up for his weekend of training!"

Mike and Rosalie looked at each other puzzled, concern creeping in. His sense of honor would have compelled him to report for duty. Where could they begin to look? They were turned away when they attempted to file a missing person report. Appealing to the State Police, Rosalie remembered the photo she had taken with Noah talking to the taxi driver. With that link, the police were able to search the dispatch records and learned that Noah had rented a car and driven west to Kalispell, Montana.

Glacier National Park Law Enforcement was painfully honest, explaining that each year three to four people disappear in Glacier and are never found.

In Kalispell, Mike talked with Pat Walsh, a detective with the Flathead County Sheriff Department. He began putting up posters regarding Noah's disappearance, asking for information. He talked with a reporter for the local television station. Nothing! No one reported seeing their son. Upon Mike's return, Caleb urged his parents to look through Noah's footlockers stored in their basement for any clue to his intentions. In his belongings, Rosalie found an itinerary.

From Kalispell, Noah planned to enter the Bob Marshall Wilderness, south of Glacier National Park. He would walk a 50-mile route to Spotted Bear Ranger Station, turn east along the northern bank of the Spotted Bear River, hiking up to the Blue Lakes. When spring arrived, his parents booked a flight to Kalispell to begin their search. Months had passed since Noah had disappeared but they hoped he remained alive.

The Park Service had cleared fallen timber that lay across the access road. At the trailhead, Mike and Caleb, Noah's youngest brother, pulled on their packs, along with a chain saw and radio provided by the Forest Service and began the hike toward Blue Lakes. The harsh winter had dropped trees across the trail, they crossed one ridge after another. They found the scat of grizzly bear and moose along the trail. Each carried bear spray but felt a deep sense of exposure in the wild back country. When they reached the campground, the men did not find any sign of Noah or his possessions. As they returned to their vehicle, Mike was exhausted, Caleb discouraged.

"Caleb, I don't know where Noah is but God knows and he is sovereign," Mike reminded his son.

As Rosalie waited for them at the Ranger Station, she talked with a young seasonal ranger who had passed a hiker along the road to the Ranger Station who look like Noah. Another ranger recalled finding Noah asleep in the middle of the trail as he was patrolling on horseback. He mentioned a family he met from Great Falls, Montana on the same trail.

Returning to Kalispell, the Pippins posted flyers and pre-recorded interviews with a local KPAX -TV host. About two later, Vern Kersey was sitting a doctor's office and picked up a flyer from a couple looking for their son, Noah. Vern, his wife, Donelle and their two children enjoyed hiking in the Bob Marshall Wilderness, They had a story to tell the Pippins. Noah had walked into their camp near the foot of an escarpment called the Chinese Wall.

At an elevation of 7,200 feet, this 1,300 foot-high limestone escarpment runs 22 miles on a north-south alignment along the Continental Divide. A heavy winter storm was moving in and the Kerseys urged Noah to turn back with them rather than venturing further. Noah politely refused.

Darnelle suggested that he could help eat the food they had brought along as they did not wish to carry it back. No, he was determined to push upward toward a saddle, the trek taking him off the trail. Once across the saddle he would drop to a lower elevation below the snow line. He told Darnelle that if caught by the snow, he would take shelter under a tree. Undeterred by their concern, he politely took leave of the family, heading up the trail.

In the fall of 2011, the Sheriff formed a search party and obtained permission to ferry supplies in by helicopter, an unusual request granted by the Forest Service as motorized vehicles and helicopters are not allowed in wilderness areas.

The team gathered and the sheriff instructed, "Think about the training this man received as a Marine. As the snow began to fall, where would he look for shelter?"

That morning, Vern had sensed a Chinook wind cresting the Chinese Wall as he prepared breakfast. As with the day Noah disappeared, the storm intervened, turning back the search party. That evening, they met a young man and his father on the trail. The young man was a diabetic and his insulin pump had failed. They were desperate to evacuate as the young man could die from the lack of insulin. The helicopter lifted the young man to safety.

The following year, 2012, the Sheriff organized a second search party. Caleb and members of the Border Patrol joined the team. Verne Kersey once again took his annual week of vacation to join the search as well and the group entered the Bob Marshall Wilderness.

Gathering at their campsite, one man called for a moment of prayer. Then, the search party split into teams, walking a parallel course through the rough terrain. Verne Kersey opted for a lower elevation as his team followed the base of the Chinese Wall. Like many limestone escarpments, sharp rocks lay along its base and

Verne was concerned for his dog's paws. They checked every hollow and fold in the terrain for a glimpse of something out of place in the scree field at the foot of the Chinese Wall.

The team noticed a huge boulder that had caved away from the limestone cliff. Upon reaching the boulder, Verne edged around the corner and a spot of color caught his attention. Climbing closer, he realized he was staring at a patch of fabric. Closer yet, he recognized the remains of a pack, a sleeping bag and canteen. There was Noah's name patch on his camouflage jacket. A photo of Noah with his mom lay atop the bag, undisturbed. Thumbing the button on his radio, he called the other search teams.

"I've found him. I've found Noah!"

They had searched for all of two hours. Verne sank to the ground, sobbing. He pulled out his phone and called Darnelle. In disbelief, she heard him say, "We found him!"

Noah's death was deemed a suicide, however, before he had departed on his walkabout, he had attempted to call Caleb. Every experienced hiker knows to arrange a search in advance if one does not return. Authorities considered the items found in Noah's pack, how he sought to take shelter in a depression under the boulder and the items he left in his footlockers. Ultimately, they concluded he had not intended to end his life in the wilderness. A new County Coronor agreed, the cause of death was changed to unknown. Noah was looking toward the future. His one flaw seems to have been that he could not accept help.

This unwillingness to accept help affects many of our veterans. They struggle to admit they cannot find their way out of the remnants of trauma and their training. The options the men use to escape their pain vary, most destructive. Men caught in a vicious cycle choose what seems to be an end to the pain, leaving their families to grieve and the men of the 1/5 to quietly acknowledge the same struggle in their own lives.

As we gathered the second evening, the Marines would honor these men just as they honor those killed in combat. Their families are designated as Families of the Fallen, indicating that the struggle for their loved ones is akin to combat within their being.

Chapter 11

Dining with Robert Irvine

The second evening of this reunion, nearly 800 men and their wives gathered for dinner. Sitting together over a meal may create a sense of unity and foster familiarity. After entering the hall, we were led to a section set aside for the Gold Star Families and Families of the Fallen. Around us, men greeted each other after an absence stretching years. The volume of conversation rose.

Charles Williams stepped to the microphone to introduce the evening's activities and welcome all those attending. Within minutes, he introduced the Gold Star families. Men and women rose to their feet, their applause drowning out all other sound. As the applause went on, I felt a sea of eyes on us. Glancing around, other Gold Star families and those who had lost their sons after the return from Iraq sat with many staring down at their plates.

One couple was struggling with the recent loss of their son and now, her father was hours away from death. Their faces creased in grief, tears slid down their cheeks. The Marines meant well, hardly suspecting the struggle for so many of us who had already walked through grief, exposed to the public eye.

Gold Star mom, Gina Trovillion, believes that the applause is for our sons and their sacrifice. My thoughts drifted back to the young man who first boarded the bus for the MCRD in San Diego. His journey brought us to this place.

Charles Williams next introduced Major General Fred Padilla and other distinguished guests before calling Robert Irvine to the podium. In introducing him, he thanked the Robert Irvine Foundation for sponsoring the reunion. The Foundation's financial support made this evening possible. Irvine strode to the podium to speak of his own history with Britain's Royal Marines and his support for the veterans before dashing back to the kitchen.

For those who are not familiar with Chef Robert Irvine, a few minutes watching Restaurant Impossible on the Food Channel gives viewers a quick introduction to his 'take no prisoners' attitude. Irvine began his career at age 15 in the Royal Marines learning to cook for large numbers of hungry men. He would later work as executive chef in the cruise ship industry. He came to public attention when he starred in the series Restaurant Impossible as a chef who reclaims a failing restaurant by updating the menu and decor along with reconciling disagreements among owners and staff. At times, the confrontations between chef and owner can be intense. As the star of the show, Irvine managed to succeed in each challenge.

Along with the success in the Food Network, Irvine never lost his affection for the men and women who serve in the Armed Forces, whether in Britain or the United States. He formed a foundation to support these men and women. The Board of the 1/5 appealed to the Foundation, seeking support for the reunion on the 20th anniversary of the invasion of Iraq. In response, the Foundation agreed to sponsor the 2022 reunion, including a formal dinner with Irvine and his team of line cooks and chefs. Steam tables with large metal containers awaited service around the room, offering salads, turkey meatloaf, mac and cheese, Yardbyrd chicken (a favorite of Irvine), and other selections with a large prime rib carved at the end of the serving line.

When Williams returned to the microphone, he introduced a veteran with a prosthetic leg. I recognized the man as the one charging up 1st Sergeant's Hill. After his brief comments, a second speaker moved toward the podium. This was the man with hooks on the end of his arms who had ridden the bus with our group the previous day.

Michael Schlitz was born in Illinois. He admits to being an adrenaline junkie. When younger, he thought he might become a policeman or fireman but eventually his desire turned to military service. After graduating from high school at age 19, he knew he wasn't ready to attend college. He enlisted in the US Army. He thought he would do a three-year stint in the Army and then head off to college. He ended up serving in the Army for 14 years.

In Sept. 2001, Michael was stationed in Korea as he watched video of the Twin Towers fall. He made the decision to go into the Army Rangers. He deployed in 2006 with the 10th Mountain division. This division was split between three missions. His first assignment was with the infantry division clearing roads of IEDs. The squad had three types of huge machines, including 'the buffalo with a large claw'. These vehicles were heavily armored. In 2006, the action was getting very violent and the vehicles were getting hit hard. As the mechanics struggled to keep up with repairing each damaged vehicle, the commander shifted his humvee over to Michael, a platoon sergeant.

On February 27, 2007 Michael and his crew rolled out just before dawn on a road-clearing mission. Their convoy included five vehicles. They rolled into a dead-end road, forcing the convoy to turn back. Unknown to them, an IED was buried below the road surface but the detection equipment could not distinguish between an IED and a metal culvert under the road. The IED consisted of 2 artillery rounds with a full propane tank.

As Michael's humvee passed over the IED, an insurgent triggered his device and the IED exploded, filling the humvee with flying shrapnel. The medic and gunner were killed instantly while the driver was trapped in the burning wreck, unable to escape. Michael's door flew open in the explosion and he was ejected, his body on fire.

As he regained consciousness, his training kicked in. He looked around expecting an ambush. Staggering upward, he moved toward the burning humvee, thinking he had to rescue his men. As he moved, he felt the flames rising around his face and realized he was on fire. Dropping to the ground, he rolled to stifle the flames. His clothing was soaked in propane and with each roll, he re-ignited. As intense pain racked his body, his muscles locked up and he lay face down in the dirt.

"I'm going to die," he thought as his brain roared at the pain.

He heard men yelling as they sprayed his body with a fire extinguisher. Relief rolled over him, knowing he was not alone. His men raced to evacuate Michael to advanced medical care. All of thirty minutes passed between the impact of the IED to the moment he rolled into surgery. Over the next eight hours, his body sufficiently stabilized to be evacuated to Germany. During the flight, his heart rate flat lined twice and he had to be revived.

From Germany, he was evacuated to a burn unit in Texas with second and third degree burns over 85 per cent of his body. He lost his ears, his nose, the sight in one eye and both hands. Only his feet remain undamaged, protected by his heavy boots. In Texas, every bed was filled in the burn unit. The doctors mercifully kept Michael in a medically-induced coma. The treatment for such severe burns involves abrading or removing the dead skin as it begins to sluff from the body. The treatment is agonizing as nerve endings are exposed. The process must be performed twice a day. With such a high percentage over most of his body, he might not have withstood the tremendous pain without going into shock.

As he emerged from the coma, he was heavily drugged. His mother and brother stood by his bedside and in his heavily-medicated state, he kept trying to send them out on patrol. They did not comply with Sgt. Shlitz' orders!

He had been told his hands were gone but as is common, phantom pain from the missing appendages made it hard to accept their absence. No mirrors were allowed in his room. The day came when he made his way to the bathroom and caught a glimpse of his appearance. He began to understand the damage to his body.

That evening, my daughter hurried after him to request that she have her photo taken with him. As they stood together, she told him how much she admired him as she has taken care of burn patients as a nurse in the hospital. She sees his altered appearance as witness to his determination to rise above the explosion that altered his life so significantly.

Like every survivor, he has gone through grief, through anger and depression. And this is why he now stands at podiums across the country telling the story of how he was injured and his will to survive. That evening, every man hung on his account, the room silent except for Michael's voice.

He retired from military service in 2010 and works for the Gary Sinese Foundation, encouraging veterans in their struggle to build a life after service to their country. He ended his talk that evening, saying, "I have a calling to give back, to use my voice to encourage others."

His story is one of perseverance against tremendous damage to his body. I sat, stunned as he finished speaking. How could one look at the bald, scarred scalp, the hooks, his limp and not feel humbled by this man's spirit to survive and surge forward in life?

As General Padilla rose to speak, Michael was a tough act to follow. Soon, Williams rose to conclude the program and the music began. Several children moved out onto the dance floor, free from their parents' urging to wait just a moment more.

Robert Irvine had assured the men of the 1/5 that this would not be the last they would see of his Foundation as his intent is to continue support once he works with a Battalion. The demand is ever growing. The following year, the Foundation's assistance would be limited. This is easy to understand when one considers the scope of their efforts.

As we turned in that night, we looked forward to one more event the following day, Easter 2023, the men and their families would gather on the beach of Camp Pendleton for a service conducted by Chaplain Carey Cash. The chaplain who had traveled every step with them as they surged into Iraq with OIF 1. It seemed fitting that he would close the reunion.

I spent a few minutes online reviewing the Robert Irvine Foundation through Charity Navigator. The organization rates a charity in three areas. I first looked at the expense ratio for the areas of Administration, Fundraising and Programs. What percentage of the capital they received did the Foundation spend in each of these areas? This is important. Any organization that spends above 20 percent on administration and fundraising deserves careful consideration before contributing financial support.

In 2022, the most recent year available for publication, the Robert Irvine Foundation reports spending 6% of revenue for administration and 3% to raise funds in support of their organization. The remaining 91% was funneled into programs under the Foundation umbrella. This is a very good percentage!

The Foundation focuses on four specific areas of service including Food, Wellness, Community and Financial Support with each operating in multiple areas. Their focus is on supporting service members, veterans, first responders and their families.

Chapter 12

Gold Star

Underlying the stories of the men of the 1/5 were memories of those who had not returned with the Battalion. A parent or partner who loses a son or husband is given the designation of a Gold Star family.

Wednesday afternoon, April 20, 2005 -- I returned home from work around 4:30 and settled down to write for an hour. Around 5:15 the back door opened and my husband strolled in. He just doesn't come home this early. We laughed over his suggestion that he return to work and moved into the living area to chat about the day.

Glancing out the window, he muttered, "oh, no."

I jumped up to find two uniformed Marines crossing our front porch. In that moment, I knew, anyone would know that grief and loss had come to our home. The Marines do not send an officer to tell you your son or daughter is injured. I understood our son was gone.

Thousands of families across the United States have faced that moment in our conflicts in Iraq and Afghanistan. In terms of numbers, we have lost far fewer than the number who died in World War II. But the grief is still bitter for the family who opens the door to the Marine officer who unexpectedly appears, his face grim.

The 1/5 lost 29 men through their three deployments into Iraq, beginning with Lt. Shane Childers on the opening day of the invasion. We first met the men of the 1/5 as they returned from the third deployment in October 2005. The Battalion marched onto the parade deck in formation, the cadence called by an officer in lockstep with his men. A few words of ceremony and the men were released to their families.

Later that evening, we went out to dinner with Alpha Company to celebrate their return. One other Gold Star family joined the assembly. We sat at the end of one table, silent, observing the men as they ate and conversed, the noise level steadily climbing. As Eric Young noted, the time in Iraq was accentuated by noise. The Marines seemed to yell, even in conversation.

No one approached as we ate dinner and sat through the speeches. As the event drew to an end, the Company commander rose and instructed each of the men to speak to us, thanking us for our son's service. They filed past, solemn and uncomfortable, muttering, "Sorry for your loss, ma'am."

I was confused. Why were they so hesitant to speak to us? Years later, I learned that many of the men were struggling with survivor's guilt. Each man was struggling to process, to grieve over their lost comrades. That evening, the men were celebrating their survival, even as many pushed aside their sense of survivor's guilt. Eighteen years later, a handful of Gold Star families attended the reunion.

"Where are the other families?" I asked one of the Marines who had been in touch with us prior to the reunion.

"Some of them couldn't make it due to other obligations," he replied. "Some of them still blame us for their son's death."

"What? How could they blame you? Doesn't the responsibility fall with those you were fighting?"

"They believe that we should have somehow saved them."

I think back to the stories I had heard from the Marines, of the ambushes, the snipers, the fire fights and the explosions of IED's. How was it possible to anticipate every single move of an enemy as they used subversive tactics on their home ground and the civilian population as shields?

For many of these families, the grief remains. They may fear that attending a reunion would drag them back to a dark place. Their loss remains deeply personal, as they hold close the memories of raising a boy to be a man and how each lost their son. Even as we talk about our experience, I was struck by the resilience displayed by many of the mothers as they related the details of their loss. Resilience acknowledges the loss but does not bow under it.

A resilient woman moves forward, stretching, bending and seeking new opportunities. She does not allow grief to become her identity but finds a way to commemorate her son or husband and to find reward in her effort.

For my husband and I, our faith was the key to living with grief. Due to my faith in God, I have the hope that I will see my son again some day in the presence of God. Not every family shares that hope and are lost in grief. Wives and children, mothers and fathers, we each grieve in our own way even as we share the loss.

For a wife, losing a husband can be devastating. The one she promised to love, with who she may have created new life, will never return. She will never wrap herself around her lover again or turn to him when she faces a difficult problem at home. She suddenly feels very alone.

Years ago I talked with a widow from the Vietnam era. Her husband was lost when his plane was shot down over North Vietnam. At the time, the government allotted a $12,000 policy to the men killed in action between 1964 to 1975. Twelve thousand to rebuild their lives and to support their children.

"I didn't have time to grieve," she told me. "I had to find a job and figure out where we were going to live. I quickly remarried. There was no time to grieve."

Years later, when she walked into the service for my son, she stopped, looking at the casket. Turning quickly, she told her current husband, "I think we had better sit in the overflow." The memories were not far below the surface.

The loss of a son is different for a parent. The mother has witnessed her waistline growing, she feel the first brush of the baby's movement. The parents raise that child, knowing that as an adult he will leave to establish his own home. They do not expect to bury their child. With the loss of a son, the mother is reminded of her creation in the intimacy with her husband and the need for comfort reminds her she was the one who gave her son life. Grief partners with love.

For those who live in small towns, who lost their sons early in the conflict, the weeks following their son's death seem like a transgress in their grief. Countless people have approached me, saying,

"I don't know how you do this. I couldn't handle losing a child."

Before his enlistment, as Marty and I discussed the possibility of his death, he would ask, "Mom, if you had to choose between my death in a head-long collision with a drunk driver or dying in service to my country, what would you prefer?"

"Are those my only two choices?" I asked indignantly. "What about choosing to live?"

We have been lauded for the sacrifice of our sons in service to our country. We had little choice in their decision to place their lives on the line when they enlisted in the Marine Corps. Even for those who survive, their lives are never the same after they return. They have given their health and opportunity on behalf of their country. Their families bear that burden with them, often at a high cost.

Gina Trovillion Patton describes her son, Tyler, as an energetic kid with a quiet sense of humor. One of five children, by the time he reached junior high, he displayed a sense of responsibility. In trying out for the school band, the band director chose Tyler for the drums in the percussion sections due to his self control.

As her oldest son, Gina and Tyler grew close as he moved through his years in school. After graduation, he began attending the Community College in his town but found he longed for something more. He chose to join the Marine Corps, viewing the Marines as the tough guys. When Gina asked him if he had prayed about this decision, he assured his mom he had. She let him go as he signed his enlistment papers. After completing Boot Camp and Infantry School, Tyler was assigned to Alpha Company, Second Platoon. Months later, Alpha Company was en route to Okinawa before deploying to Iraq.

Before his deployment, Tyler loved hanging out with his younger siblings. Now they missed him! He sent them small packages, each a surprise that they eagerly anticipated.

In Fallujah, his squad noted Tyler was meticulous with an eye for small details that others did not notice. Something small would catch his attention and he would warn the other men to be alert. Due to this alertness, he found caches of weapons and spotted buried IEDs. His attentiveness helped to save the lives of the men around him.

The battle for Ramadi in one sense was their toughest deployment as the conflict had descended into guerilla-style fighting, the patrols never knowing when shots would ring out or an IED would explode as they passed along their route. They struggled with moments of sheer boredom while on watch, playing word games even as their eyes scanned the area around them. The boredom was broken by deprivation in forward positions, often without adequate water and fire fights leaving them diving for cover as they formulated a plan to confront the enemy concealed in neighboring buildings.

On June 15, 2005 an IED under a vehicle carrying five Marines took Tyler's life. Hours later, three uniformed Marines appeared at Gina's door to inform her of Tyler's death. She stood there, her mind questioning what had happen. She was not angry. In disbelief, she said, "I don't understand?'

The news of Tyler's death spread quickly through the town she lived in. The story was covered by the local news station and the community gathered around her. That Sunday, her pastor rose to address his congregation, asking them to attend the funeral out of respect for Tyler's service to our country. He led the service, a responsibility that he often left to an associate pastor. Over 500 people from the church and surrounding community attended the service.

"To see over 500 people attend the funeral impacted me. I could see that Tyler's life and service mattered to so many people. This was a comfort to me.

"My greatest support came from the military community. Well, I should say, from the Second Platoon. These were the men Tyler served with and knew best. The men of the Second Platoon were instrumental in my healing," she says.

The men of Second Platoon have continued to check on her and share their stories of Tyler. Now and then she receives a letter telling her how much Tyler meant to the author. Those moments are like tiny sparks of light and warmth that illumine her daily routine.

For most Gold Star moms, months, even years pass before their vision clears. As we rise out of grief, we begin slowly to see our loss from a different perspective. Gina searched for a way to honor her son. The Casualty Assistance Officer (CAO), suggested that she take her grandchildren out for ice cream. Gina searched for an idea with special significance for Tyler. As Tyler's birthday approached, she asked a grandson if he had ever sampled German chocolate cake, a favorite that she baked for Tyler, at his request, every year on his birthday. She was excited to learn that her grandson had never sampled German chocolate cake and a new tradition was born.

"That year, on Tyler's birthday, I went out and purchased a humongous, over-priced piece of German chocolate cake. My grandson and I sat in a park as he savored the treat and we talked about Tyler. Her grandson knows the uncle he never met though Gina's memories. This began a quiet tradition celebrating Tyler, the son she believes she will see one day in the presence of God.

Gina thanks God for the 23 years she shared with her eldest son. Her wish, as she remembers Tyler, is for her actions to grow out of her love for her children to bless other people. She feels "a calling to walk down lonely roads and through difficult times" with Marines struggling to adjust today.

Resilience! Every one of the women I interviewed showed an amazing resilience toward deep loss. I was amazed at how many of the Gold Star mothers I met shared a common faith. I wondered about those who did not appear at two consecutive reunions. Did they remain bitter? Does faith allow those who grieve to move on more easily? I'm not sure I want to answer that question after witnessing the loss that weighs down some of the Gold Star families.

Many Gold Star families struggled as the United States pulled out of Iraq and Afghanistan. Did the collapse in these countries minimized their son's sacrifice? The death of their son or husband seemed to accomplish nothing. Many express grief through anger, struggling to keep the anger from turning destructive. One mom, Shellie Starr becomes discouraged when she sees how quickly we have forgotten the men who have given their lives on behalf of our country.

"We say we will never forget," she says. "What about the men who fought in Vietnam? In twenty years, who will remember the men who fought in Iraq? I hate seeing what is happening in our country. I've become very skeptical."

She is referring to the political upheaval of recent years and the turning away from traditional values. Her grief for her son becomes mingled with her grief for the decline she see in this nation.

Shellie and Brian Starr lost their son, Jeffrey, on Memorial Day, 2005. Jeffrey was born as the second of three children, sandwiched between two sisters who both excelled in school. He often made his mom laugh, a happy-go-lucky kid who excelled in B and C level work in school. As the only boy in the family, at times he felt as if he was competing with his sisters who were both high achievers. He told his mom, he wanted to be cool and have fun. Shellie visited the school counselor to express her concern about his performance. The counselor laughed at her, pointing to his grades as proof that he was doing just fine compared to many of the other students.

As he approached the sixth grade, Shellie and Brian were becoming concerned about his choice of friends so she chose to home-school Jeffrey for a year while he regained his footing and established his own identity.

"I thought I would continue the next year but Jeffrey made it clear that he wanted to return to school."

"Mom, I don't have any friends," he complained.

"You have me," she said. Jeffrey rolled his eyes.

"Exactly! Mom, I don't want you. I want to hang out with friends."

His comment stung but Shellie understood and re-enrolled Jeffrey in the public school system. After the year home-schooling, his old friends had moved on. With three schools feeding into the Junior High, Jeffrey quickly found new friends that seemed to lend a better influence.

As graduation approached, Jeff thought the Marines were the best of the Armed Forces and would be good training for pursing a career in law enforcement. He would enlist, serve four years, gain experience and move on to the Police Academy. As he moved toward the end of his four years of service, he had begun to consider all sorts of ideas about what he wanted to do.

Shellie and Brian remember the Marines as being good for their son. He quickly adapted to the requirements and structure of the Corps. He served through three deployments in Iraq, first through the invasion, then in Fallujah and Ramadi. Shellie watched as her son bloomed, becoming a man who was respected for his leadership. She had not seen this side of Jeffrey in his early years. For a kid who just wanted to have fun, he showed a good bit of upright character. He inspired other men to follow him.

Jeffrey was first assigned to Bravo Company as the 1/5 prepared to enter Iraq. Living together, training as a Platoon, the men bond over MRE's and challenges in physical training. When he returned home for a visit, he brought Kenn Boles with him. The two had formed a tight friendship. Shellie and Brian accepted Kenn as part of their family and appreciated Kenn's mechanical knowledge.

As the men deployed to Iraq with his third deployment, Jeffrey had been promoted to Corporal. The men in his squad followed his leadership as he led the patrols and held them accountable in the fire fights. On May 30, 2005, taking the point position, he led the patrol as they paced the streets of Ramadi. As the tallest man in the squad, he stood above the other men, making him an excellent target for a sniper.

Jeffrey lost his life the day before he was scheduled to leave Iraq and return to the United States. When Kenn learned that Jeffrey had been killed, he wondered how he could enjoy life when his friend and fellow Marine would not live to see another day? How could he face the Starr family?

Stateside, on Memorial Day, Shellie returned to their home after shopping with her daughter. Brian remained at work. As she pulled into the long driveway, she noted a van sitting in front of their home. Pulling closer, she noted the license plate and then the Marine Corps sticker. Leaping from the car, she began to scream as her daughter stood frozen, uncertain how to react. With her screams, the officer approached, attempting to make his announcement.

He requested that she call Brian, asking him to come home. He instructed her not to tell Brian the reason for her request. Even today, this strikes me as odd. They did not want him to receive the news over the phone but what was he to think as he was summoned? Together they called their eldest daughter who was studying for a degree as a physician.

Shellie called her pastor but found herself comforting him. Many of the Gold Star moms recall sharing that experience in becoming the comforter in the loss of their child. The moment is surreal as the words pour from your mouth even as your brain questions why you are not the one receiving comfort. Somehow, we come through.

Shellie had come to faith at an early age. After the funeral, she was not sure she could remain in a relationship with the God she had known throughout her life. Had she not been faithful? With Brian, they raised their family in church on Sundays and in church-related activities through the week. How could God take her son, her only son?

There is an irony to that question, as Christians hold the belief that God offered his only son in our place for what we have done wrong. I have heard well-meaning callers compare this to offering our sons to our nation's service. Through the previous two years, Shellie had prayed for Jeffrey, believing that God would protect him. And now what? He was gone. Shellie believes, as does Gina, that she will see her son again in the presence of God after her death. But that does not diminish the grief at his loss.

Grief and sorrow walk hand in hand. I am reminded of a quote by essayist Christian Wiman, "When we are alone, even joy is, in a way, sorrow's flower: lovely, necessary, sustaining, but blooming in loneliness, rooted in grief."

Anger is one part of grieving the loss of someone we love and may be projected in different directions. Some parents are angry at their son's fellow Marines. These men fight as a team, for the man on their left, the man on their right. Why could his fellow Marines not have protected him? Others are angry with the government or their neighbors for comments that seem less than gracious.

My anger chose my neighbor's son, a drug-addled, pathetic, lying piece of humanity or so it seemed. Why had God chosen to take my son and left this other young man, a parasite on the human race? I could hardly believe that in time this man would find redemptionas he did.

Strong words? Yes, but my thoughts reflected the anger I chose to direct at this boy for my own mental health. Some parents even struggle with anger against their own child even as they are repelled at betraying their love. In time, Shellie wrestled her questions to the ground and renewed her relationship with the God she had known for so many years.

In the first year, after Jeffrey's death, the Starrs did not hear from Kenn Boles, a man they had welcomed into their home. Kenn sank into anger and depression, using alcohol to numb the pain of loss over the next four years. He simply could not visit the Starr family, feeling that he had let them down.

Then one day, he found the courage to pick up the phone to call Jeff's family. They remembered him and accepted his request to visit. Together they visited Jeffrey's grave and spent hours talking about the time in Iraq. He told them stories of their service together. Brian and Shellie listened, without judgement. Kenn felt as if he had returned home. He was becoming whole again.

Listening has become the Starr's way of connecting with the Marines of Bravo Company, the men who once fought alongside Jeffrey. They are happy to pick up the phone when one calls to check in and they listen.

"Our mission is to support the Marines of the 1/5," she says.

Shellie isn't one to travel but Brian loves meeting with the men of the 1/5 at the reunions. He quietly listens to the stories, the memories that fly between the men.

Are the Gold Star families welcomed at the reunions? Are we a reminder of those men lost? I imagine we are but the men of the 1/5 have made us welcome twenty years after they first set foot on Iraqi soil. They tell the Gold Stars, "you are part of the 1/5 family now. We stand with you."

At my son's funeral, I met a young Marine. When Marty died, he qualified to be buried at Arlington National Cemetery but I chose to hold him close. He is buried in his home town. As the crowd dispersed at the cemetery following the service, I noted a small group of Marines who had served as pallbearers standing to one side. I walked toward them to thank them for their service as pallbearers. One man met me halfway.

"I want to thank you for honoring my son today." I began. "Thank you for taking the time to travel here to honor him."

"Ma'am, I am the Marine who escorted your son back to the United States from Iraq," he replied.

I was caught off guard, his comments unexpected. A lump rose in my throat and tears rose to my eyes.

"Thank you," I said, my voice breaking. "Thank you for taking care of my boy. Thank you for seeing him home safely."

"Ma'am, I was happy to serve. I'm sorry for your loss and pray that you will find comfort."

Turning, I pointed toward the crowd.

"Do you see that man in the sweater?" I asked. "That's Marty's dad. Please go talk to him, tell him you brought Marty home. He will want to talk with you as well. I so appreciate you telling me that you were the escort for my son."

He turned, walking toward Marty's dad who thanked him profusely. Two days later, we received a call from a good friend whose father had attended a church service in southern Texas. He spoke of a young man who had risen during the meeting to speak of a funeral that he had just attended in line with his official duties. This was the Marine that had escorted our son home.

"As I serve in this assignment, I have been screamed at, spit on, kicked and had doors slammed in my face," he told those gathered. "This is the first time I have been thanked for escorting one of our Marines home who was killed in action."

Not all Marines carry this sense of honor and duty in their service to the Corps. That day, I was proud of the Marine I met and how he carried himself. With that in mind, I watched the men of the 1/5 mingle through the opening reception at the reunion nearly 20 years later. I thought of honor and survivor's guilt as I sat, my head bowed through the applause honoring the Gold Star families and their sons. They were the ones who had served under fire, knowing that in a moment, their lives could end with a bullet or explosion. Out of pride in their service, pride as Marines, they chose to honor us and our sons.

Chapter 13

More Than a Sunday

My bare feet sifted the cool sand as a chill breeze swept through the crowd slumped on folding chairs on Camp Pendleton's beach. Sunday morning, we gathered for an Easter Service. A white stole lay draped over the chaplain's camouflage uniform. Looking out over the crowd, Chaplain Carey Cash greeted those gathered. His acolyte stood to one side with a guitar.

Twenty years earlier, the chaplain had traveled north with the 1/5 as they crossed the border into Iraq. In his book, *A Table in the Presence*, Chaplain Cash makes note of the close relationship between faith and men who move into combat. His presence was essential to the survival of these men, a confidant in remembering their accomplishments.

More than a remembrance, for families who have lost a loved one, the Easter Service speaks of the hope of the resurrection. The families gathering that day savored hope as they remembered those gone before them into eternity. Easter is the day we celebrate redemption from all we have done wrong, the forgiveness of God and the promise of life with God. The promise of resurrection is hope for the men who have lived through combat and given up their comrades.

As the voices of those gathered lifted in a hymn, the notes seemed puny against the vast space around us, just as they must have seemed years earlier in the vastness of the Kuwaiti desert. Chaplain Cash read from his Book of Prayer as the group followed his words. As the service drew to a close, Chaplain Cash invited those of the Christian faith to come forward to share a sip of juice, a portion of bread to celebrate the living body of Christ.

I watched as men and women fell into line in front of Chaplain Cash. His acolyte waited for a second line to form. Moving forward, I shared the deep respect the men and their families held for their chaplain who once rolled forward beside them into conflict.

When my son returned from his second deployment, I asked him about the ambush that occurred as the men entered Bagdad.

"Oh, it was no big deal, mom. I sat at an intersection for several hours."

"I don't think so," I said. "You were caught in crossfire as you entered the city. The fighting was intense! You weren't just sitting at an intersection.

"How do you know about that?" he stuttered.

"Busted!" Laughing, I waved a copy of Cash's book at him. "I know the story. You want to tell me about it?" He studied the cover.

"Mom, this is by Chaplain Cash! Do you know who he is? He's our chaplain. He is such a great guy." His words tumbled out.

"Yes, we caught an interview with him after your first deployment. His story was heart stopping! I've actually ordered several copies for your grandparents and others in the family."

Carey is a man who understands the Marines without being repelled by their rough approach to the sacred. As he made his way through the 1/5, he made God visible for men who were questioning their future as the invasion approached. Going into battle, he knew the Marines feared they might be injured or captured, even killed

One asked, "Chaplain, are there certain sins that God can never forgive?"

Chaplain Cash recounts how he assured the young Marine, saying, "God's love never changes. If we're willing to open up our lives to his healing power, there is no sin he won't forgive."

The Marine was still worried, "You don't understand, Chaplain. How could God ever forgive me when He and I both know what I've done?"

The chaplain again assured the Marine that he could explore God's love with him through Jesus.

Another Marine turned the question over, asking, "Chaplain, what if someone in your life has truly hurt you ... deeply? How do you ever forgive them? You're always talking about forgiveness. You don't understand what . . . what they did to me."

Seeing the man's pain, the chaplain assured him, "Forgiveness is not easy. Sometimes, forgiving is more for our good than for the sake of those who have hurt you. Do you want to talk about what happened to you?" *

Chaplain Cash went on to talk with the men he counseled, speaking of "broken homes, betrayal, failures, and disappointments, dying marriages, expectations not met and debilitating fears."

These men met a chaplain who would listen, who understood the pain they kept hidden beneath a tough exterior. He found a way to meet them over the issues in their lives that remained unsettled and questions about the future. After the Marines returned from Iraq, I marveled at how these men were eager to seek out the chaplain if he were stationed near their duty location. He had earned their respect in sharing those months of deprivation and pain.

Cash was raised in a traditional family, his mom remained at home with the children as his dad flew missions as a fighter pilot in a F-4J fighter over North Vietnam. His father was shot down over the Gulf of Tonkin and barely survived due to the action of a rescue swimmer who cut him free from the parachute shroud that threatened to drag him below the surface of the ocean. He grew up watching his parents' marriage based, as he said, on "a seamless union of respect, love and mutual support."

As he prepared to graduate from high school, Cash received a full scholarship from The Citadel to play football while completing his undergraduate degree. Two months later, he married a young lady he met in high school. Charity knew what it meant to be married to an officer in the Navy as her father was a Naval Chaplain.

In 1993, just a year after they had married, Cash was diagnosed with a tumor on his brain stem which had caused blurred vision and headaches. The doctors determined that the tumor was not cancerous but could not provide an accurate prognosis. Eventually, he consulted a world-renown neurologist who had diagnosed and treated hundreds of similar cases. The neurologist assured him that

*A Table in the Presence, Carey H. Cash, W. Publishing Group, 2004

he could lead a normal life though he could not guarantee the tumor would not grow larger. The surgeon was also willing to write a letter certifying that Cash could serve in the Armed Forces. Shortly after this diagnosis, his symptoms disappeared, not to return. The Navy accepted him into service, providing he serve under a medical waiver.

His father-in-law suggested that he consider becoming a chaplain and Cash enrolled in a Theological Seminary. After graduation and several years of working as a pastor in a small church, he enlisted as a chaplain and was assigned to the 1/5 in 2001. Shortly after his arrival, two planes struck the Twin Towers in New York. Cash understood he would be serving with a battalion headed into combat.

As he considered the likelihood that he would serve in Iraq with the 1/5, a verse came to mind, a verse from the Bible given to a young woman as she considered her longevity if she should appear before a despotic king.

"And who knows but that you have come to this position for such a time as this?"

He understood that this was what God had called him to do and he believed that God held his life secure whatever might come. Even as he assured young Marines of God's love, he also worked to deepen their knowledge of God.

On March 20, 2003, 1/5 was ordered to move across the border into Iraq as the United States went to war against the forces of Saddam Hussein. The moment weighed heavily on each group of 25 Marines as they waited in their amphibious assault vehicles to begin rolling. The chaplain moved from vehicle to vehicle assuring these men that God was rolling with them. He spoke of God's angels, messengers of fire, that would accompany them as they rolled in pursuit of justice for a people who lived under tyranny. As the moment of departure drew close he uttered a quick prayer and stepped into his own humvee, accompanied by Second Class Petty Officer Redor Rufo, who was assigned to assist and protect the chaplain. *

* Per the Geneva Convention, a Chaplain does not carry a weapon.

Twenty days later, Chaplain Cash would search the faces of these men who had survived after fighting for nine hours as they entered Bagdad. Non-combatants had been pulled back as the officers suspected that the fighting would be heavy. They could not have guessed the hail of bullets and rocket-propelled grenades that would descend on the 1/5 as they trailed through the streets of Bagdad en route to one of Hussein's palaces.

As Cash entered the palace grounds, he did not know what to expect from the men. The first day of the invasion, the 1/5 had lost 2nd Lt. Shane Childers and the chaplain shared the grief and shock of the men over the death of one the men they loved. Now, they had lost Gunny Jeffrey Bohr, another noncommissioned officer the men highly respected. As helicopters dropped from the sky to evacuate the wounded, the chaplain searched the faces of the men who gathered in the courtyard.

Looking around he found "deep and quiet joy" in their eyes as they nursed their wounds and rested from the combat. He spoke to one man, then another, asking how they were doing. He listened to their celebration over the protection he had promised God would provide.

As the 1/5 returned from Iraq, the chaplain was assigned to a new duty station. Now, 20 years later he had the opportunity check on the men he counseled and consoled in those anxiety-fraught days. I looked up to see chaplain moving through the crowd. In the days following our son's death, I had spoken with Chaplain Cash, thanking him for his kindness and his work with our son. I told him how much Marty appreciated his ministry to Alpha Company. Now, I thanked him again as he met us amidst the reception, offering a quick hug and a few words.

Not every Marine finds a successful path to life beyond the Marine Corps. What makes one man different from another? What causes a veteran to spin into a downward cycle while another finds success in his recovery.

My thoughts return to the faith of many of the men in Alpha Company. In the following chapter I relate the story of Dave Cecil, who served two tours in Iraq. He told me he found God in the bottom of a fox hole. He was born to Christian parents but until he began his own relationship with God, his faith made little difference.

Was faith a key to successful recovery? If so, then what of Joshua Kegley who struggled with an addiction to alcohol? He had also been raised in a Christian home, professed his faith and sung with a team in his church. Yet, he had fallen to the degradation of alcohol, destroying his liver and kidneys. Why was faith not enough for him to surge forward, conquering his addiction and find success in life? I don't believe it is enough to credit the pull of alcohol on the body as I know others who have successfully fought their addiction and found fulfilling lives. Was it the trauma of what they experienced in Iraq? Then, what of Kenn Boles who climbed off a bar stool to race motorcycles at Bonneville and fight raging forest fires?

Kenn came out of a dysfunctional home. He did not feel that home was a place he wished to return when he left the Marine Corps. A number of Marines will tell stories of homes and dysfunctional families. Some men went on to establish successful relationships and raise well-behaved children, completely unlike what they had known in their younger years.

Social workers suggest that a genetic heritage might be a factor in whether an individual is susceptible to addiction. The condition is labeled as Alcohol Use Disorder (AUD). A number of factors contribute to a man being susceptible to alcohol abuse and effects 29 million people in the US each year. However, studies show that many of those retaining a genetic preference for AUD do not become alcoholics. Instead, contributing factors, including a history of trauma, a lack of coping mechanisms, unemployment and regular exposure to alcohol, may draw a man into dependency on alcohol. All of these factors seem to be a part of what the men of the 1/5 experienced within the Corps from 2003 to 2005 as they fought in Iraq. * Families of those lost would argue that these factors do not apply to their sons. However, when scientists dig a little deeper, they often find that neurological damage along with the memories of what a man has endured will not allow that man to heal. Alcohol becomes the means by which a man subdues the

*https://www.healthline.com/health/alcoholism/is-alcoholism-genetic

trauma and the struggle to live day to day. After talking with the Marines as well as medical personnel, I have come to believe that the power of the supernatural can effect a healing when a man truly opens his spirit.

This is a multi-prong approach. The healing process is often painful and difficult, requiring that a man journey back in time to confront the damage that has been done while simultaneously seeking a healthier life-style and pursuing a life of faith.

There is something to be said for faith as a catalyst toward resolving the trauma these men endure. I found the story of Dave Cecil, his energy and the story of being a reborn spirit interesting. Could this approach be a means of redemption for the men who have moved through trauma, through injury and the loss of the lives they once knew? Could there be hope in the faith that the chaplain shared with his men?

Chapter 14

Homeward Bound

As the Easter service wound down to a close, the crowd began to drift toward a block building. At the edge of the sand, one Marine was the center of attention and I drifted closer to hear what he was saying that so absorbed those who listened.

Dave Cecil had served with Alpha Company through their months of combat in Bagdad and with H&S in Fallujah. With 1/5's deployment in Ramadi he provided support stateside. He described the lasting effects of the traumatic brain injury he had suffered in combat and how the injury effected him 20 years later. In chapter four, I wrote about post traumatic stress disorder. Traumatic brain injuries are distinctly different from PTSD in veterans though the symptoms between the two are sometimes confused.

April 2001, Cecil was selling commercial insurance when he noticed a recruiting poster for the Marine Corps. He was struck with the desire to serve his country. He walked into the recruiting office a month later, never dreaming that in four months two planes would strike the Twin Towers and his country would enter a conflict far from his home. Mentally, he was not prepared for the rigors of boot camp and found the training very difficult.

"I had no exposure to the military prior to enlistment. When I graduated from boot camp, the world was a very different place than when I enlisted. I looked up at the crowd watching our graduation, feeling how the support for the military had grown in just a few months," he recalls.

"I thought the School of Infantry would be easier than boot

camp but the training got harder. One day, watching the rain pour off the eves of the building, I wondered if I was going to get through this. I knew I had to take this one day at a time, one month at a time."

Assigned to 1/5, he encountered General Fred Padilla and grew to respect the training the battalion commanding officer expected.

"He was preparing us for Iraq! I believe that the training, though tough, prepared us for what we would face in combat. General Padilla stood out for his ability to stay calm and to make decisions when fighting was at its worst."

Two years later, when the third deployment arrived, Cecil was scheduled to be discharged. When the 1/5 departed for Iraq, he did not deploy with Alpha Company as his contract was nearly finished. Remaining stateside as the rear guard, he cared for the details as the 1/5 staged. He sought the items they needed, arranged transportation when a man returned and made life easier for the Marines. Cecil drove the wounded men who arrived to the hospital. He recalls one wounded man, adjusting to jet lag and moving from Iraq to the harried pace of southern California, who expressed is relief in not having to work out transportation. Cecil accomplished his duties even as his brain struggled with the aftermath of being on the receiving end of a detonated rocket. Ironically, one of the Marines mounted the debris of the rocket on the front of his vehicle while in Fallujah as a message to the Iraqis of the Marines' durability.

After his discharge, he earned a degree in Interpersonal and Organizational Communication - the man rapidly spits out words a like fully-engaged machine gun! Ten years after he left the Marine Corps, he was experiencing memory loss, confusion, and headaches. He was diagnosed with post traumatic syndrome and given a disability rating. When he began losing his sight, the medical personnel found that an arachnoid cyst was causing the temporary loss of sight. He questioned what more his injuries would do.

Symptoms of a traumatic brain injury (TBI) may include memory loss, headaches, seizures, dizziness, vision changes, paralysis or being out of balance, and mood swings. Deteriorating language skills, cognitive impairments and the ability to problem solve may also begin to appear. Those who seek treatment soon after the injury have a better change of recovery.

As Cecil began to recognize some of these symptoms, the VA made an appointment with the ophthalmologist. When the doctor asked him if he had experienced an explosion, Cecil asked, "which one?" Tests revealed that he had an image or impression of the explosion on the lens in his eye and as a result a black cataract was forming. With surgery, a new lense was implanted.

Sadly, many of the Marines experience traumatic brain injuries (TBI) due to the number of explosions they experienced in Iraq. Many experienced concussions while for others, like Cecil, the damage was more extreme. With TBI, the damage is discovered as symptoms appear over the years. Prompt treatment would have limited the damage. The Marines are not alone in experiencing TBI as statistics show that 5.2 million people are diagnosed with TBI and each individual exhibits their symptoms in a unique pattern. This makes diagnosis difficult which Cecil discovered when seeking treatment through the Veterans Administration. He was frustrated when the answers did not come immediately but this inspired him to begin his own education in what he was experiencing. Even as he sought treatment, his marriage disintegrated. The battle over custody of the couple's children turned bitter.

After his lens replacement, a doctor suggested they do a brain scan and the test showed part of his skull as being deformed. The MRI was very painful as the machine uses powerful magnets. These magnets interact with any metal remaining in the body. Cecil felt as if metal was trying to force its way out of his skull as the magnets spun. He remembered inspecting his helmet upon discharge. He found the helmet was damaged from the impact of the rocket.

At this point, he determined, "I need help!"

When he walked into the VA Center for an appointment, one of the personal walked up to him and shook his hand.

"Congratulations! You've been healed!"

Cecil was quite certain that he was not whole. He was not going to settle for a diagnosis of PTSD and began his own research into TBI. He refused to allow his injuries to hinder his pursuit of what interested him. He believes research showing the brain has plasticity and can in part build new pathways as we step up to new challenges.

Physical therapy, rehabilitation psychology, occupational and speech therapy all contribute to improving one's outcome. One therapist I spoke with uses exercises with pencil and paper along with games of Skipbo following the Feuerstein method. For those requiring care givers, financial assistance with training and respite care may be essential.

I thought back to the Corpsman who insisted that PTSD be used as a stepping stone to reach for the next level of achievement. Cecil was a walking-talking example of such effort.

Today, Cecil interacts with researchers studying the effects of TBI on the brain and has written a book describing his pursuit for answers in living with his injuries. He is going to live with confusion, irritation and memory loss the rest of his life. He choose to see his injuries as a blessing which have allowed him to enjoy his children as they grow and develop. He remains determined to make the most of his new reality and to enjoy life. He is excited for what researchers will learn in the future. Through therapy he has learned to recognize how others are living and how he is different. He compares this new reality to trying to move around a city with which he is not familiar without a map. Once he began to learn about traumatic brain injuries and how to live with the symptoms, his life improved dramatically.

Reflecting this attitude, he says, "When I have the truth, I find it starts breaking out in other places in my life." He believes that one reward for his effort is a healthier relationship with his children.

"When we lose something, we gain something much healthier. The injury took that away but in its place, God gave me so much more. I find God has better plans. My heart is so fulfilled. I feel complete."

Identity - Purpose - Direction. Everyone of us seek these three values as we find our place in our culture. The man who offers four years to the United States Marine Corps is given these three values.

He is a Marine, one of the few and the proud. He understands he is in the Marine Corps to serve his country and is taught to follow orders, all part of his purpose and direction. Not every Marine finds a successful path to life beyond the Marine Corps.

Sitting in an airport, watching people flow past me as I wait for a flight, I see a young man striding along the corridor past each gate. Have you noticed such a man? His shoulders are unbowed, his head up, his eyes sweeping back and forth across the crowds. His stride is one of purpose and he rarely smiles. His posture and alert surveillance scream military, possibly a Marine en route to his next assignment.

He has a purpose as he strides forward. He is following the orders he has received. Get here, be here, be the proud, the few, the Marine. He doesn't have to wonder what he is to accomplish today, that has already been decided by men of higher rank. Sure there are times, when the going is slow but ultimately, he will be on time at the correct location.

When a man is discharged from the Marines, he suddenly has no sense of where he might be headed other than the home he grew up in, if that home still exists. He knows his purpose has changed. He no longer answers to the Armed Forces of the United States. He is free to make his own choices and unfortunately, those choices are not always clear.

As Josh Shores has written in his memoir, "I was trained to kill. There is little demand for a trained killer in our society."

When discharged, a Marine examines the possibilities of employment and where he might find a home, a place to lay his heart. For many, a career in law enforcement is attractive for the common elements in handling weapons and maintaining civil order. Law Enforcement is structured with a chain in command. The training in self denial shifts easily from one arena to the next. Up to 20 percent of law enforcement may have military experience as part of their background.

Akin to a career in law enforcement is pursuing a position as a private security guard either through a local company or in service to the wealthy and privileged. Ryan Ackermann accepted a position in personal security and grew to dislike the position. In contrast,

Sgt Major Luke Converse provides security in his hometown hospital and has defined the job as one of service. Josh Shores and Kenn Boles each entered fire fighting as a secondary career. Both domestic and wildland fire fighting require men to maintain their physical fitness, constantly adjusting to the plan of attack and feeds their desire for an adrenalin rush. Imagine urgently digging stumps, roots and brush in the shadow of two-story flames roaring toward you.

Cyber security has become a lucrative field for Marine Corps veterans as the threat to our country moves online. IT Security is growing exponentially. Corporations have found that men skilled in working as a team have a valuable skill set is valuable in human resources or sales work. Both require the ability to communicate well and set goals.

One area few think of when examining careers beyond the military is the mechanical side of aviation. For those with mechanical abilities, positions as an aircraft mechanic can be high pressured but well paid. Many Marines cushion their reentry into civilian life by enlisting in the Reserves. This helps extend the benefits they receive as veterans.

One of the most intriguing employment opportunities for military men lies in becoming a military contractor though there are some drawbacks. This is not necessarily steady employment as contracts are often for three to six months. A military contractor may decide to pull out of a region without warning, leaving men to search for other opportunities.

Wade Spence left Iraq and the Corps after the deployment to Ramadi. Four years later, he chose to return to Iraq.

"I left the Marine Corps with a honorable discharge and enrolled in George Washington University, earning a degree in International Relations. I believe that school kept me from going into a downward spiral. After graduation, I was unwilling to admit I had PTSD as I wanted to work in security. In 2008-2009, I thought I was above PTSD. I was active, suicide was not an option for me. I wondered if I should re-enlist.

"I chose to go back to Iraq as an independent contractor. I spent hours thinking about this. Three reasons worked into my decision to return. I had loved traveling in my early years with my parents. I learned through drinking alcohol that I have an addictive personality. And finally, I loved the adrenalin rush.

"In some ways, life is simpler in Iraq. We live in the present. The stress left me feeling alive. I put the PTSD in a closet, locked the door and went forward. Sometimes, the PTSD would escape and I would have a few bad days until I could get it back under control. Working overseas, I traveled the world. As a contractor, I would work for nine weeks and then have three weeks off. I was able to see so many parts of the world."

Veterans are considered highly desirable as private contractors due to their training. The Rand Corporation reports that in 2008 private contractors numbered 163,000 in Iraq and Afghanistan. By 2013, this number had dropped to a mere 6,000. Their employment frees the men of the Armed Forces from being assigned to non-combat positions in logistics, food service, maintenance and security. Currently the number of private contractors consists of 70 percent American citizens, 20 to 30 per cent from other nations and at one time up to 10 percent from Iraq.

The United States Army instituted a program titled Logistics Civil Augmentation to coordinate the work of private contractors. One sticking point for the governments of other nations is that the United States has not signed onto the International Code of Conduct. The Code of Conduct has been constructed through the International Criminal Court. On the surface, this would seem to be an ethical step but as international experience has shown, abiding by a code of conduct drawn up by nations opposed to the United States has been used unjustly at times and against our best interests. Such an imbalance should be considered by any Marine considering employment as a private contractor.

The compensation for a private contractor can be as low as $25 an hour for entry level, rising to $50 an hour for mid level contractors and up to $300,000 annually for the highest senior levels with both combat and management experience. Benefits are rarely offered as part of the employment package and the rate of injury or death can be higher than average for US citizens in regions of high risk.

While seeking employment in law enforcement, firefighting, security or as a military contractor, many men returning from Iraq have not fully dealt with the trauma they have experienced or the post traumatic stress that began within a month of their return. The question arises whether stepping so quickly from military combat to law enforcement is detrimental to both the man and the people he dealt with in his community? Did he struggle more adapting to life with his family as the demands of his occupation extended the flow of stress and cortisol through his brain? Becoming a member of a squad with the police department does not necessarily equate to the squad of a Marine Platoon as the training does not prepare the men equally for what they encounter. For a man who is seeking to re-establish his identity and purpose, law enforcement may fall short and he struggles with managing his interaction with a civilian population.

Many of the Marines recognize that they do not want to live in the past but struggle to find the right niche. Wade Spence summarizes their desire to remain active and useful.

"Right now, I'm figuring out what the next step will be. I'm enjoying traveling throughout our beautiful country. I need to find a new purpose. I don't want to be stuck in a cubicle and I need to plan financially for the future. I volunteer like my dad who set an example for me. He worked with kids in a baseball program. Someday, the multiple sclerosis I've developed may leave me unable to get around. I don't want to work for the next 20 years and then when I retire, not be able to be active, wishing I had done what I enjoyed earlier in life.

"It is important to tell our stories. People need to know the sacrifices we made in serving our country. They need to know what sacrifice is."

Ultimately, every individual is looking for a place where he belongs, a place he feels at home. For the Marines of the 1/5 who served in years 2003 through 2005, the reunions offer a place where they do not have to explain the past. They are among those who lived the same experiences. This is akin to anyone who has lived in a sub-culture of our society. We have a quiet set of shortcuts that in conversation do not require us to explain ourselves. We begin to feel at home even though we may be far from the place we were born.

In a sense, returning to the theme of Easter morning, Marines have successfully made the transition from battlefield to civilian life. In a sense, they are birthed in the violence of training and combat; reborn when they return and successfully make the transition. One man stood on the pavement outside the reunion of 2023. He continuously fed a lit cigarette to his lower lip, as he clutched the leash of the service dog. Minutes earlier, I had noted his presence inside as he talked with other Marines.

He stepped outside, looking for that sense of stillness that allowed him to re-group. He regained his balance and after few minutes returned to the reunion.

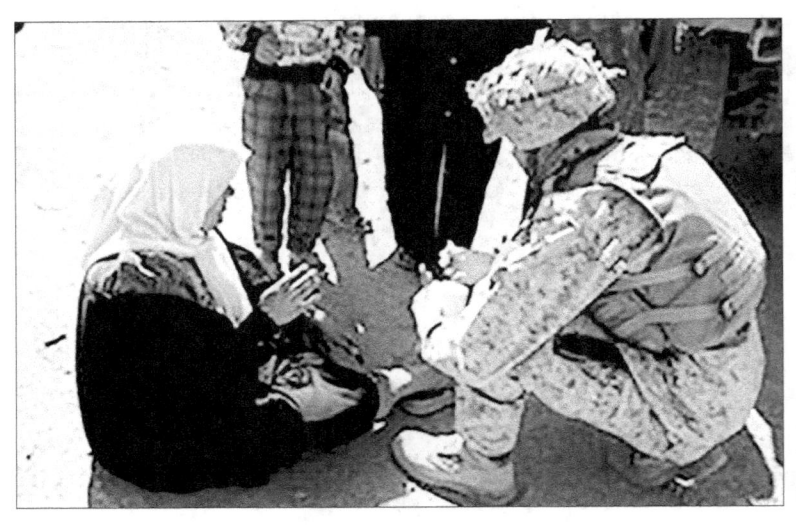

Chapter 15

Reflections

The first draft of this manuscript was completed in September 2024 as our nation marked the twenty-third anniversary of the destruction inflicted by terrorists within our shores. In a somber moment, I reflected on the violence the men of the 1/5 had inflicted on Iraq. I thought of the wounding and deaths of members of our armed forces. We recognize the aftermath of our struggle that began on September 11, 2001 reaching into Afghanistan in 2021 and beyond.

Many of the young men sent into Iraq enlisted either right before or right after 9/11. They were deeply affected by the events of that day and understood that we would be a nation at war. That somber day in September led to the death of my son and the sons of other Gold Star families. The Gold Star moms of the 1/5 quietly note this anniversary as our memories drift back to those we sent into the US Marine Corps.

In the days after Marty was killed, I kept thinking about the sacrifice of other families who have lost their sons in traffic accidents, through illness or due to genetic defects. I thought of those stripped of their loved ones by the drugs that flood over our southern border. How was I any different in my grief than those parents? We each know loss. Each of us has our own struggle as we contend with what threatens our lives. I do not believe the death of my son assigned me to a special pedestal. Yet, in a small town culture in 2005, the media, friends, even complete strangers flocked to us, elevating our sacrifice of our son.

Some went so far as to compare the loss of a son to the sacrifice that God made in sending Jesus. One woman told me, "Your son gave his life for me."

I found that a stretch. She was under no immediate threat from those on foreign shores but I understood that she regarded this as a fight to be free from the evil that has crept across our planet. We have long offered the lives of our sons in the protection and liberation of other nations. My son did not regard himself as some sort of hero.

Repeatedly he told me, "Mom, I have a job to do."

Nothing more, nothing less.

In one sense the price paid by the Marines in surviving the conflict in Iraq and Afghanistan has been unending. In that struggle, those who survived have the more difficult task. How do they choose to live after their guns fall silent and they return to a society mostly at peace?

As we entered the conflict in Iraq, we did not study the history or recall the lessons of early explorers in this region. Our diplomats followed the time-honored tradition of trying to establish a central leader with whom they could negotiate and rebuild the government.

They might have read the recollections of Gertrude Bell, a British explorer and diplomat in the late 1800s who traveled extensively throughout the Arabian Peninsula and the region that is now the country of Iraq. She found the region rife with tribal and clan rivalries who for centuries fought each other and schemed for power. Each served only their own interest and betrayal was considered clever.

With Operation Iraqi Freedom (OIF), we entered Iraq with good intentions, to restore justice to a people who lived under tyranny. Ultimately, our men returned disillusioned, many believing the destruction they had committed, the violence they had endured had failed to accomplish a just end. We had failed to resolve the problem of evil that lies at the base of every struggle for power

It is easy to cast blame on the machinations of the politicians but the foundation of failure is much broader than a set of negotiations. By our very nature, each of us covets what we do not possess. Our leaders, once in power, seek greater authority and abuse the advantage they have been given when in office. Within the desire for personal gain, evil finds a toehold. What the Marines encountered in Iraq was the end result of men who began with just a toehold. Evil increases, expands, grows until it is checked by sacrifice. We cannot hope to resolve the desire for power and authority until we address the issue of evil.

In his memoir Josh Shores says he repeatedly found himself slipping into behavior he would have found unconscionable a year earlier. As his platoon fought ruthless men day after day, they began to regard the violence as acceptable. The restraint of this Marine's upbringing in a peaceable society kept slipping away. Looking back at his actions, he regrets the attitude that brought him to the point of living by animal instincts, reflecting a moral injury. Out of that regret, he believes that we must learn to talk to each other to resolve our conflicts, rather than descending into the grim instinct to kill or be killed by men who fight for their very survival. Was this not the impetus for the United Nations?

Man has long advocated for resolving our conflicts with talking to each other and yet this has failed. We do not understand the problem of evil.

Quoting Solzhenitsyn who endured years in the Soviet gulag, "The line separating good and evil passes not through states, nor between classes, nor between political parties either -- but right through every human heart -- through all human hearts. This line shifts. Inside us, it oscillates with the years. And even within hearts overwhelmed by evil, one small bridgehead of good is retained."

We may question whether a small bridgehead of respect for human life remains in a fanatic who chooses to kill indiscriminately. The problem of evil remains and cannot be resolved by mere talking. Not until the very nature of man, all men, undergoes a radical transformation and we find a means to sacrifice our greed and corruption for the good of others. This is impossible in our human realm. I have come to believe that only with supernatural transformation will conflict cease.

Until that day, the number of names on the 1/5 Memorial on the grounds of Camp Pendleton, indeed, on every memorial will continue to grow.

Until the day comes when the Prince of Peace returns, my hope and desire is to see every man that returned from the 1/5 finds the life they were willing to die for in the sand of Iraq. My hope is that they will come to rest in a place that feels like home.

If you require counseling and medication, then please avail yourself of the resources available. Reach deep within to challenge the grief that troubles your mind. Seek the medical resources that will bring relief to the physical injuries. Reach out and talk with those that have made the journey to healthy living and found peace within that passage. May you live to find joy in the rewards for which you struggled and survived.

www.ingramcontent.com/pod-product-compliance
Lightning Source LLC
Chambersburg PA
CBHW070326130626
46556CB00007B/2754

* 9 7 9 8 9 9 9 5 5 0 6 0 6 *